Bond
No.1 for exam success

SATs Skills

Arithmetic Workbook

10–11+ years
Stretch

Do not write in this book

OXFORD
UNIVERSITY PRESS

Great Clarendon Street, Oxford, OX2 6DP, United Kingdom

Oxford University Press is a department of the University of Oxford.
It furthers the University's objective of excellence in research, scholarship,
and education by publishing worldwide. Oxford is a registered trade mark
of Oxford University Press in the UK and in certain other countries.

British Library Cataloguing in Publication Data

Data available

978-0-19-274566-8

10 9 8 7 6 5 4 3 2

Paper used in the production of this book is a natural, recyclable product
made from wood grown in sustainable forests. The manufacturing process
conforms to the environmental regulations of the country of origin.

Printed in China

Acknowledgements

Cover illustration: Lo Cole

Although we have made every effort to trace and contact all copyright
holders before publication this has not be possible in all cases. If notified
the publisher will rectify any error or omissions at the earliest opportunity.

Links to third party websites are provided by Oxford in good faith and for
information only. Oxford disclaims any responsibility for the materials
contained in any third party website references in this work.

Number

> **Helpful Hint** 12
>
> Remember that on a number line, the negative numbers (those to the left of zero) appear to get larger the further below 0 you go.
>
>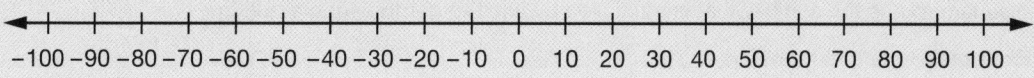
>
> −100 −90 −80 −70 −60 −50 −40 −30 −20 −10   0   10   20   30   40   50   60   70   80   90   100

Ⓐ Answer these questions.

1 Look at the arrows and write the numbers they show on the number line.

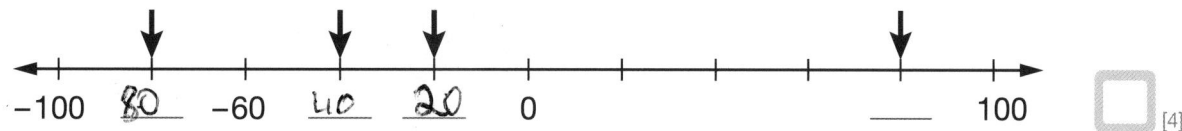

−100   *80*   −60   *40*   *20*   0   ____   100   [4]

2 Look at the arrows and write the numbers they show on the number line.

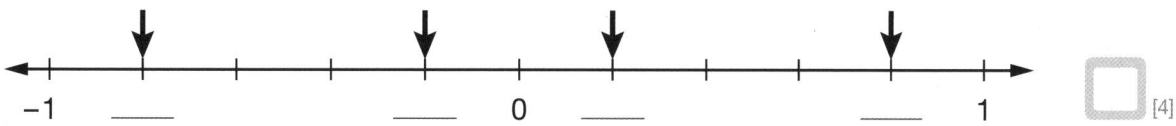

−1   ____   ____   0   ____   ____   1   [4]

3 Write these numbers in order, largest first.

  −4312   −4231   −4123   −4321   −4213

_____   _____   _____   _____   _____   [1]

4 Write these numbers in order, smallest first.

  −924.3   −9312   −923.4   −9321   −931.2

_____   _____   _____   _____   _____   [1]

5 The temperature was 28°C but dropped to −2°C at night. By how many

degrees did the temperature drop? _____ °C   [1]

6 The temperature was −17°C and rose by 23°C. What temperature did

it reach? _____°C   [1]

12

Helpful Hint

A square number is the product of a number multiplied by itself.

Example: $4 \times 4 = 16$ so 16 is a square number. $4^2 = 16$

A cube number is the product of a number multiplied by itself then multiplied by itself again.

Example: $4 \times 4 \times 4 = 64$ so 64 is a cube number. $4^3 = 64$

A square root is the number that is multiplied by itself to get the square number.

Example: Square root of 16 is 4. $\sqrt{16} = 4$

A cube root is the number that is multiplied by itself then by itself again to get the cube number.

Example: Cube root of 64 is 4. $\sqrt[3]{64} = 4$

(B) Answer these questions and show your workings out.

1 $9^2 + 10^3 =$

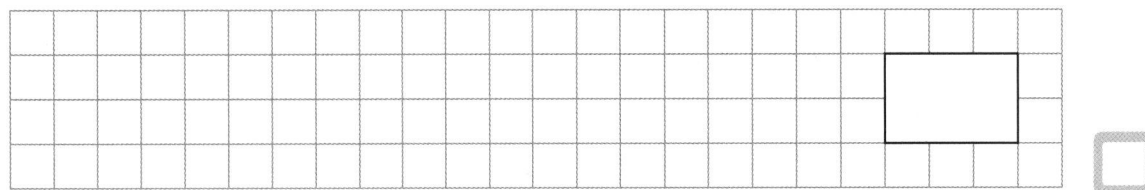

[1]

2 $12^2 - 3^3 =$

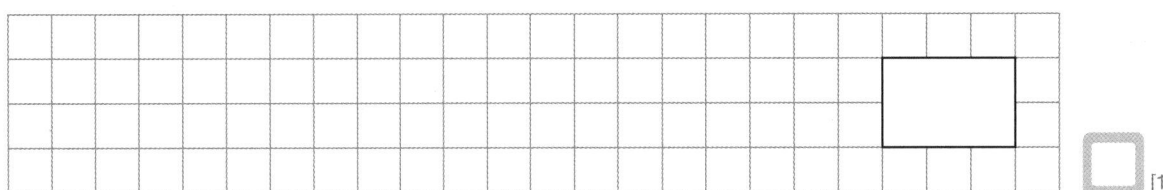

[1]

3 $5^2 \times 2^3 =$

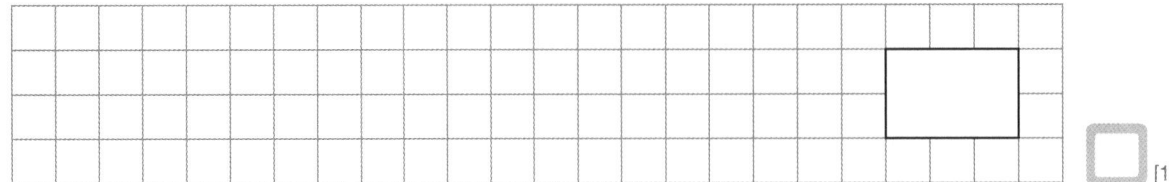

[1]

4 $6^3 \div 3^2 =$

[1]

4

Helpful Hint

Length: 10 mm = 1 cm   100 cm = 1 m   1000 m = 1 km

Weight: 1000 g = 1 kg   1000 kg = 1 tonne (t)

Volume: 10 ml = 1 cl   100 cl = 1 l   1000 ml = 1 l

Remember to change the amounts to the unit of measurement required in the answer.

ⓒ Answer these questions and show your workings out.

1   7985 g + _____ = 8 kg

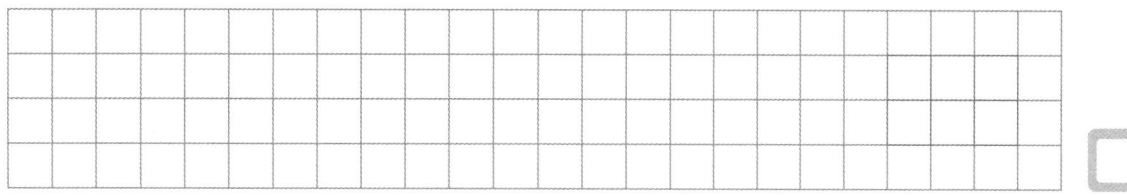

[1]

2   2.5 t + 800 kg = _____ t

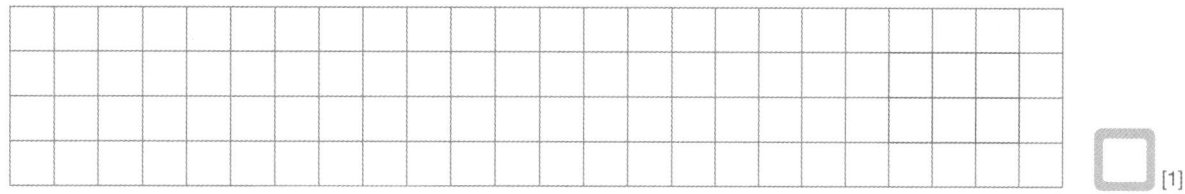

[1]

3   1600 g + _____ = 3.5 kg

[1]

4   _____ kg + 1234 g = 5 kg

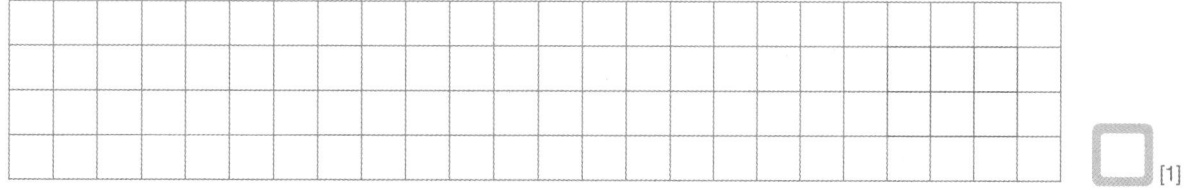

[1]

5   4500 ml + 2 cl = _____ cl

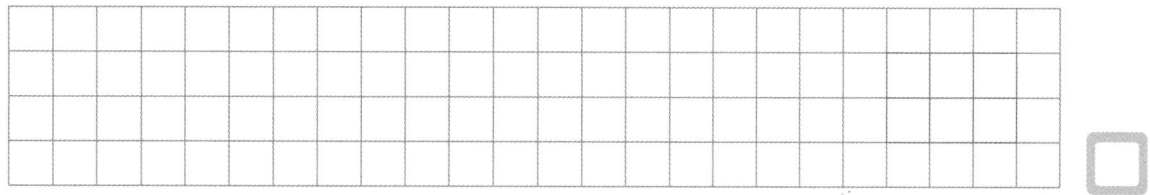

[1]

5

Unit 1

Word problems

D) Solve the word problems and show your workings out.

1 Jali makes cakes to sell at the market. He spends £83.74 on ingredients which will make 280 cakes. Jali sells each cake for 75p.

   a How many cakes does Jali need to sell before he is in profit?

   _____

   b How much profit does Jali make if he sells all 280 cakes?

   _____

   [2]

2 The temperature in Paris is 4°C warmer than London and 4°C cooler than Venice. Venice is 5°C cooler than Shanghai, which is 8°C cooler than Las Vegas which is 40°C.

   a What temperature is London? _____

   b What is the difference in temperature between Paris and Shanghai?

   _____

   [2]

3 Lukas reads 5 chapters of a book each week, which is half the number that Vladimir reads. Vladimir reads 2 more chapters a week than Marat who reads twice as many chapters as Sergei.

   a How many chapters a week does Marat read? _____

   b How many chapters a week does Sergei read? _____

   [2]

6

Multi-step questions

 Helpful Hint

BODMAS provides the order in which we solve **equations**.

Brackets, **O**rder, **D**ivide, **M**ultiply, **A**dd, **S**ubtract

Sometimes we refer to **BIDMAS**. It is the same but the word Order is replaced with the word Indices. Both words mean finding square and cube numbers, square roots and cube roots.

(A) Underline the correct answers to these questions.

1 $3(7 + 4) - 14 \div 7 =$ 1.6   2.7   23   31   [1]

2 $24 \div 3 + 6(10 - 8) =$ 1.6   2   20   132   [1]

3 $100 \div 2(4 + 6) + 15 =$ 2.9   5.75   20   515   [1]

4 $2^2 \times (5 + 3) \div \sqrt{16} =$ 1   1.4   2   8   [1]

5 $7 \times 10 - 7 + 4 \times 3 + 2 =$ 35   77   149   203   [1]

6 $15 \div 3 + 2 - 4 \div 2 =$ 1   1.5   5   7.5   [1]

(B) Answer these questions. You do not need to show your workings out.

1 $5 + 28 \times 2 - 3 =$ _____ [1]

2 $5^2 + 20 - 2^3 =$ _____ [1]

3 $25 \div (2 + 3) \times 2 =$ _____ [1]

4 $3(6 - 2) + 8 - 4^2 =$ _____ [1]

5 $175 \div 5^2 + 3^3 =$ _____ [1]

6 $13 + 2 - 6 \times 2 =$ _____ [1]

7 $15 \div 2(10 - 5) + 6 =$ _____ [1]

8 $2.5 \times 2 - 3 + \sqrt{81} =$ _____ [1]

9 $\sqrt{100} + 3(5 + 6) - 20 =$ _____ [1]

15

Helpful Hint

A **number** (or **output**) **machine** starts with a number, takes it through a sequence of operations and gives an amount as the final output.

Example: Put 2 in an output machine that follows these operations:

$$+ 2 \Rightarrow \times 10 \Rightarrow - 3$$

2 + 2 = 4, 4 × 10 = 40, 40 – 3 = 37

The output is 37.

ⓒ Put each number through the output machines and find the output.

1 ⇒ +2 ⇒ × 10 ⇒ –3 = Output

a 5 ☐   **b** 12 ☐   **c** 17.4 ☐   **d** 306 ☐   ☐ [4]

2 ⇒ ÷ 2 ⇒ + 5 ⇒ ÷ 10 = Output

a 8 ☐   **b** 40 ☐   **c** 310 ☐   **d** 685 ☐   ☐ [4]

3 ⇒ – 5 ⇒ × 2 ⇒ square = Output

a 7 ☐   **b** 10 ☐   **c** 25 ☐   **d** 30 ☐   ☐ [4]

4 ⇒ + 3 ⇒ ÷ 2 ⇒ cube = Output

a 1 ☐   **b** 5 ☐   **c** 15 ☐   **d** 17 ☐   ☐ [4]

Helpful Hint

With **multi-step problem solving**, ask yourself what you need to do to the numbers and in which order you need to do the steps. Highlighting numbers and signs can be helpful.

(D) Break each of these questions down into **TWO** steps to find the answers and show your workings out.

1 £50.00 – (71p + 95p + £4.49 + £1.18) =

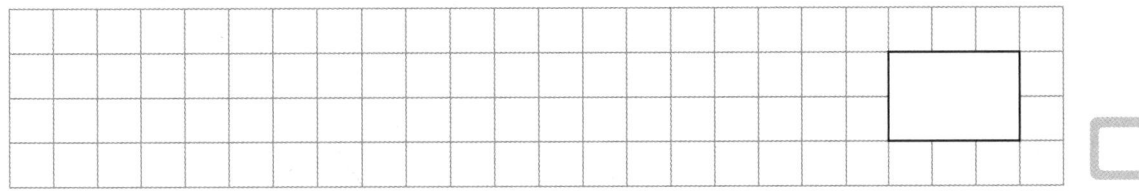

[2]

2 (14 × 30) × 48p =

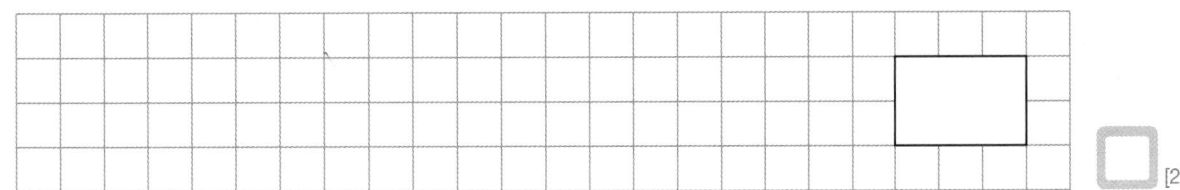

[2]

3 1m – (8.4cm + 6.5cm + 7.21cm + 5.48cm + 9.1cm + 6.3cm + 4.98cm + 5.82cm) =

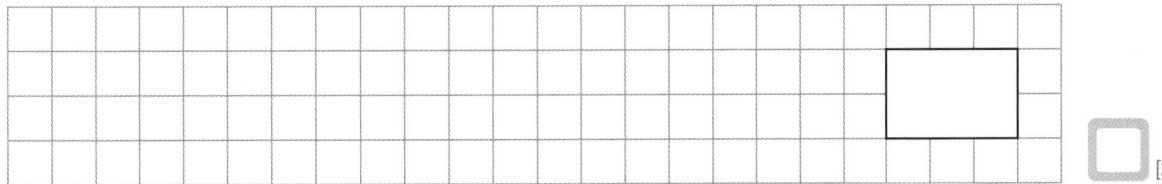

[2]

4 2kg – (246g + 129g + 972g + 43g + 27g + 15g + 7g + 1g) =

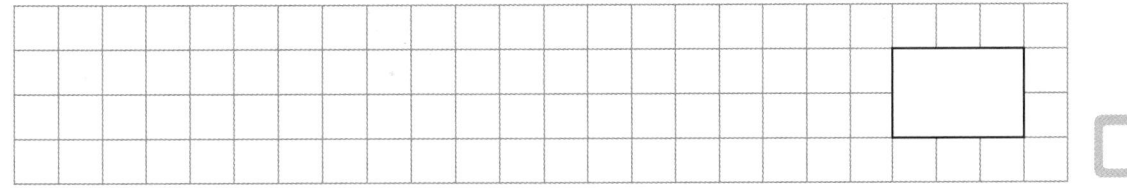

[2]

5 (£12.70 + £6.85 + £7.99 + £2.46 + £1.97 + £1.12 + 93p) ÷ 2 =

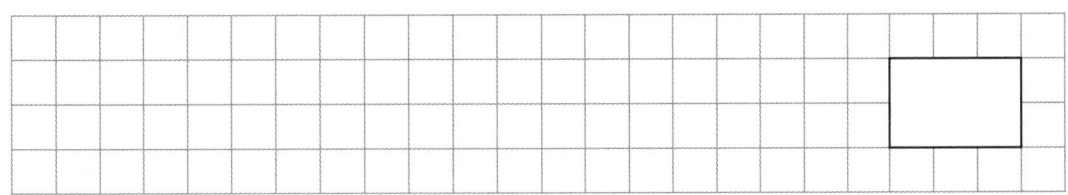

[2]

10

Unit 2

Word problems

(E) Solve the word problems and show your workings out. Write your answers on the lines.

1 Mr Nash is going to buy 12 rolls of wallpaper to decorate his bedroom. The shop has three special offers.

OFFER 1	OFFER 2	OFFER 3
Buy 2 rolls of wallpaper for £12 each and get one roll free.	Buy 6 rolls of wallpaper for £50.	All wallpaper £7.99 a roll.

a Which offer will cost Mr Nash the least amount of money? _____

b Which offer will cost Mr Nash the most amount of money? _____

c How much money can Mr Nash save if he buys the best offer? _____

[3]

2 Jacob is going to Sammy's house for a party that starts at 15:30. He is going by bus. The buses leave every fifteen minutes on the quarter hour. He has to walk for 8 minutes to his nearest bus stop but there is a bus stop right outside Sammy's house. The bus takes 25 minutes to get to Sammy's house.

a To get to the party in time, when must Jacob leave his house? _____

b If the bus home is delayed by 4 minutes, how long will Jacob's journey home be? _____

[2]

3 Mr Quinn's minibuses seat 20 people and his coaches seat 50 people. He has lots of minibuses, but only seven coaches. The local school has 620 pupils and 20 teachers. What is the smallest number of vehicles that can be hired to take them all on a school trip? _____

[1]

6

Estimating and rounding

 Helpful Hint

We can round a decimal number to a certain number of **decimal places** (d.p.) like this:

3.2576 rounded to 2 **decimal places** = 3.26 (the 7 in the thousandths column rounds the 5 up to a 6).

2.1932 rounded to 3 **decimal places** = 2.193 (the 2 in the ten-thousandths column rounds down so the 3 remains 3).

Ⓐ Answer these place value questions.

1 Underline the digit in the 10 000s place: 700 6<u>0</u>0 500   [1]

2 Underline the digit in the ten thousandths place: 193.472 3<u>0</u>0   [1]

3 Underline the digit in the millions place: 987 654 321   [1]

4 Underline the digit in the hundred thousandths place: 1.234 321 47   [1]

5 Underline the digit in the 10 000 000 place: 285 933 176 000   [1]

6 Underline the digit in the $\frac{1}{10\,000\,000}$ place: 0.283 917 465   [1]

7 Underline the digit in the thousands place: 9 753 123 495 000   [1]

8 Underline the digit in the hundredths place: 159 357.741 268   [1]

Ⓑ Round these numbers to 3 decimal places.

1 429 504.3124 = _____   [1]

2 123.145 678 9 = _____   [1]

3 958.5852 = _____   [1]

4 19 283.746 5 = _____   [1]

5 3513.824 79 = _____   [1]

6 50 607.0819 = _____   [1]

7 1742.853 96 = _____   [1]

8 0.123 456 789 = _____   [1]

16

Helpful Hint

Numbers can be **rounded** to make them easier to work with. You may have to round numbers to the nearest ten, hundred, thousand, ten thousand or hundred thousand. Remember, if the next digit to the right is 4 or less you round down. If the next digit is 5 or more you round up.

Example: 23 418 **rounded** to the nearest thousand is 23 000.

Hint: Look at the hundreds digit. (The hundreds digit is the next digit to the right of the thousands digit.)

ⓒ Round these numbers to the nearest 1000, 10 000 and 100 000:

1 957 301 _____ ☐ [3]

2 105 976.3 _____ ☐ [3]

3 858 585.12 _____ ☐ [3]

4 915 285 012 _____ ☐ [3]

5 93 827 891 _____ ☐ [3]

6 8 134 587 _____ ☐ [3]

ⓓ Round these numbers to the nearest $\frac{1}{1000}$ and $\frac{1}{100000}$.

1 0.936 285 174 _____ ☐ [2]

2 0.1 299 876 _____ ☐ [2]

3 1.628 954 731 _____ ☐ [2]

4 9.954 631 027 _____ ☐ [2]

5 0.179 494 915 _____ ☐ [2]

6 2.851 231 549 _____ ☐ [2]

30

Helpful Hint

Rounding numbers up or down to estimate an answer is helpful, especially when calculating with numbers that have lots of digits.

Example: 192.36 × 29 can be rounded to 200 × 30 to find an approximate answer of 6000.

(E) Estimate answers to these questions by rounding up or down to the nearest hundred.

1 571 × 936 ≈ _____ × _____ = _____ [1]

2 162.9 × 34 567 ≈ _____ × _____ = _____ [1]

(F) Estimate answers to these questions by rounding up or down to the nearest thousand.

1 13 549 × 81 024 ≈ _____ × _____ = _____ [1]

2 89 542 × 285 928 ≈ _____ × _____ = _____ [1]

(G) Estimate answers to these questions by rounding up or down to the nearest million.

1 1 953 764 × 8 154 237 ≈ _____ × _____ = _____ [1]

2 8 525 946 × 1 346 792 ≈ _____ × _____ = _____ [1]

(H) Estimate answers to these questions by rounding up or down to 1 decimal place.

1 7.352 × 8.063 ≈ _____ × _____ = _____ [1]

2 3.821 × 4.732 ≈ _____ × _____ = _____ [1]

8

Word problems

ⓘ Solve the word problems and show your workings out.

1 My new patio is 3.792 m long by 4.92 m wide. The stone slabs that will make up the patio are 1 metre squares and they cost £9.89 each.

   a Estimate how many stone slabs I will need: _____

   b Estimate how much my new patio will cost: _____

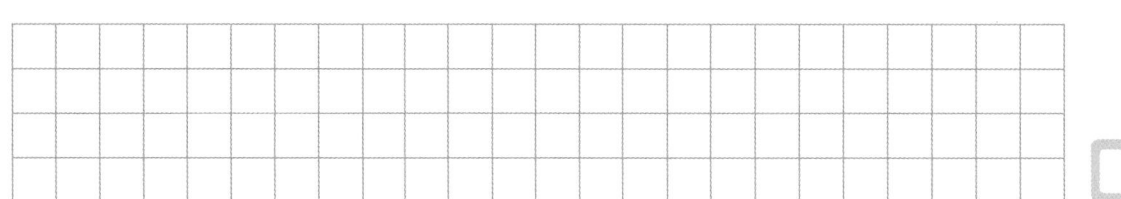

[2]

2 Raj is making a brick path using red, blue and green bricks. The red bricks are 78.6 mm long. The blue bricks are 10.39 cm long and the green bricks are 0.2 m long. Estimate how long these paths will be with the following combinations.

   a 39 red bricks + 12 blue bricks + 24 green bricks _____

   b 15 red bricks + 21 blue bricks + 9 green bricks _____

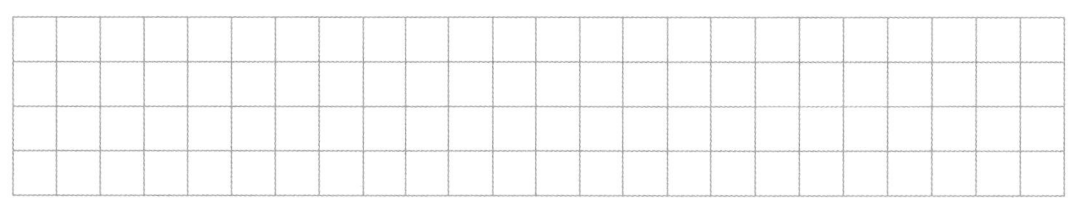

[2]

3 A school uniform shop is having a sale. Estimate how much it will cost for the following items.

Sweaters	£8.99	Shirts	£6.99
Trousers	£9.75	Skirts	£9.25

   a 1 sweater + 5 shirts + 1 trousers _____

   b 2 sweaters + 4 shirts + 2 skirts _____

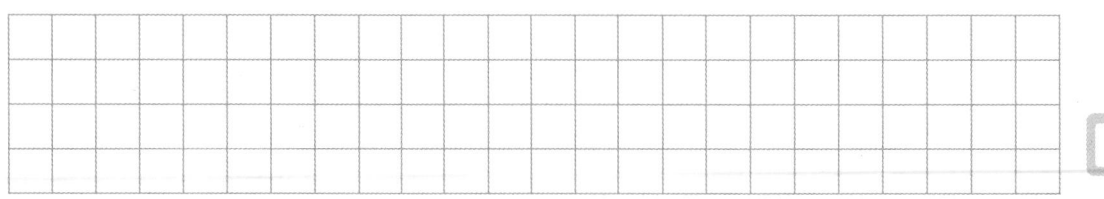

[2]

Decimals, fractions and percentages

 Helpful Hint 26

To change a fraction to a decimal number, you can divide the **numerator** by the **denominator**.

Example: $\frac{3}{8} = 3 \div 8 = 0.375$

If you have a **mixed number**, convert the fraction part into a decimal and then add on the whole number.

Example: $2\frac{3}{8}$ $\frac{3}{8} = 0.375$, so $2\frac{3}{8} = 2.375$

To change a fraction to a **percentage**, first divide the **numerator** by the **denominator** and then multiply by 100. ($100\% = \frac{100}{100} = 1$ so the value doesn't change.)

Example: $\frac{3}{8} = 3 \div 8 = 0.375$; $0.375 \times 100 = 37.5\%$

Because 'per cent' (%) means 'out of 100', to change a **percentage** to a fraction write the number as the **numerator** in a fraction and the **denominator** as 100. Then make sure the fraction is in its **simplest form**.

Example: $64\% = \frac{64}{100} = \frac{16}{25}$

To change a **percentage** to a decimal think of it as a fraction with a **denominator** of 100 and divide the **numerator** by the **denominator**.

Example: $64\% = \frac{64}{100} = 64 \div 100 = 0.64$

To change a decimal to a **percentage** multiply the decimal by 100.

Example: $0.32 = 0.32 \times 100 = 32\%$

Ⓐ Write each fraction as a decimal. Then write it as a percentage to the nearest whole number.

1 $\frac{3}{5} =$ _____ = _____ [2]

2 $\frac{19}{20} =$ _____ = _____ [2]

3 $\frac{6}{40} =$ _____ = _____ [2]

4 $\frac{8}{15} =$ _____ = _____ [2]

5 $\frac{40}{65} =$ _____ = _____ [2]

6 $2\frac{6}{8} =$ _____ = _____ [2]

7 $2\frac{5}{6} =$ _____ = _____ [2]

8 $1\frac{1}{3} =$ _____ = _____ [2]

9 $8\frac{2}{3} =$ _____ = _____ [2]

10 $1\frac{28}{40} =$ _____ = _____ [2]

Ⓑ Write each decimal or percentage as a fraction in its simplest form.

1 0.84 = _____ [1]

2 0.36 = _____ [1]

3 2.58 = _____ [1]

4 90% = _____ [1]

5 62% = _____ [1]

6 420% = _____ [1]

26

> **Helpful Hint**
>
> To find a fraction of an amount, divide the amount by the **denominator** and multiply by the **numerator**.
>
> To find a **percentage** of an amount, divide the amount by 100 (because % means 'out of 100'), then multiply by the **percentage** number. Alternatively find 10% and work out combinations from there.
>
> To find a **decimal fraction** of an amount, remember 'of' is the same as 'multiply', so simply multiply the two numbers together.
>
> Look at the fractions you are working with and make sure they in their **simplest form** before you do any multiplying or dividing. This means the numbers you have to deal with are as small as possible.

ⓒ Answer these questions as quickly as you can.

1 What is 30% of £250?

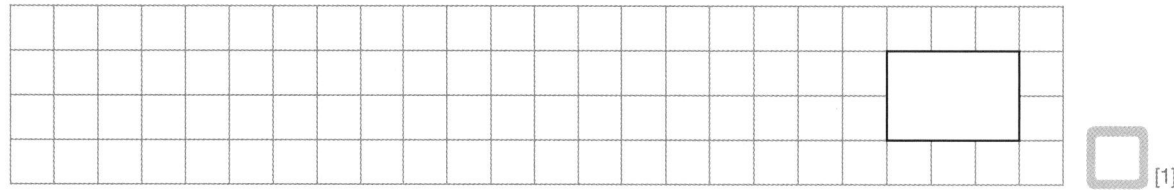

[1]

2 What is 15% of £860?

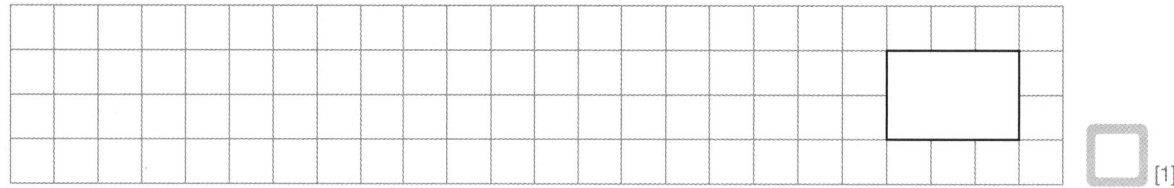

[1]

3 What is 82% of 3000?

[1]

4 What is 7% of 490?

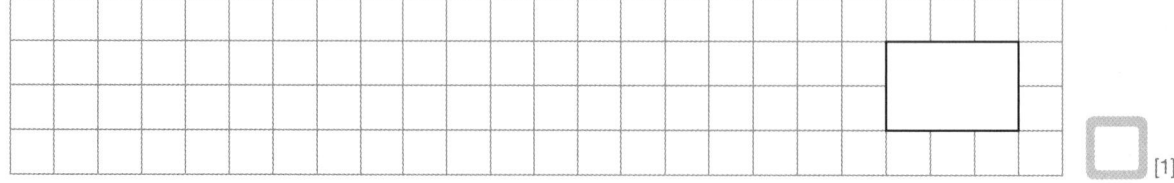

[1]

5 What is $\frac{12}{15}$ of 180?

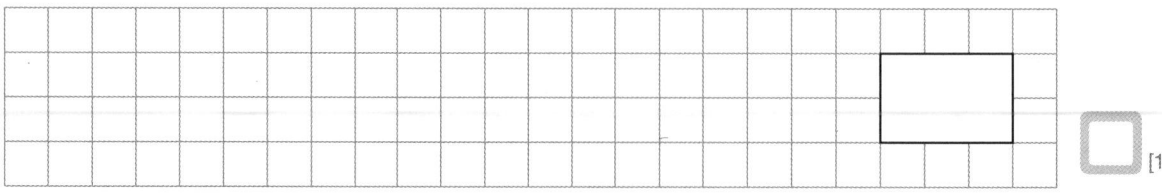

[1]

5

Helpful Hint

To find **percentage** reductions, compare the old price to the new price by writing them as a fraction. Then convert this to a **percentage** (multiply by 100). Look for chances to make simpler **equivalent fractions**.

$$\frac{\text{new price}}{\text{old price}} \times 100$$

Then subtract this answer from 100 to find the reduction. Use the example and the steps below to help you.

Example: I buy a book costing £14.40 that was originally £18. By what **percentage** had it been reduced?

Step 1 Compare the new price to the old price. $\frac{14.40}{18}$

Step 2 Write this as a **percentage**. $\frac{14.40}{18} \times 100 = \frac{2.4}{3} \times 100 = 0.8 \times 100 = 80\%$

Step 3 Find the **percentage** reduction. 100% − 80% = **20%**

Always compare with the **original** amount.

(D) Work out each percentage reduction. Show your workings out.

1 £30 reduced to £25.50 = _____

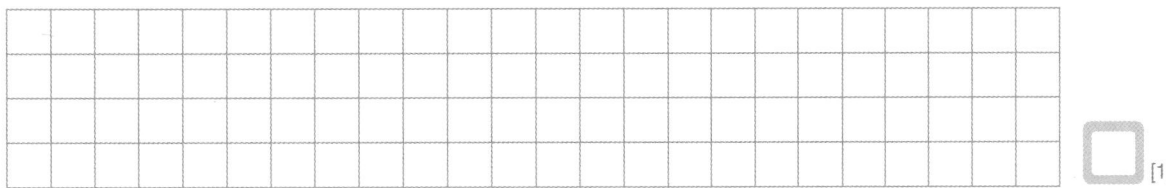

[1]

2 £120 reduced to £72 = _____

[1]

3 145 cm reduced to 94.25 cm = _____

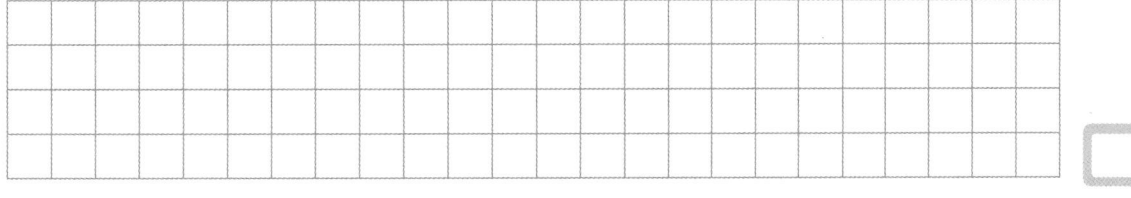

[1]

4 168 g reduced from 280 g = _____

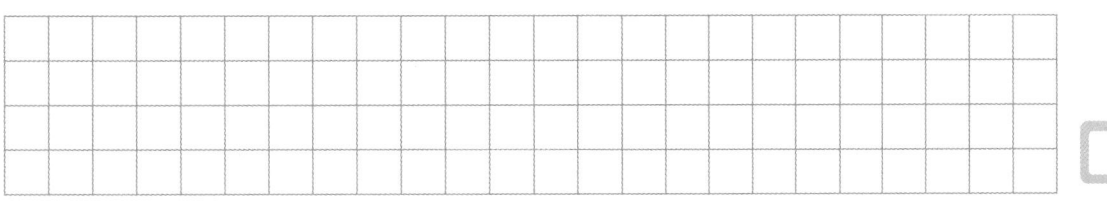

[1]

4

Unit 4

Word problems

(E) Solve the word problems and show your workings out.

1 Lily has 120 stickers. She gives Elle $\frac{2}{5}$ of the total number of stickers. She gives Jess 0.4 of the total number of stickers. How many stickers does Lily have left?

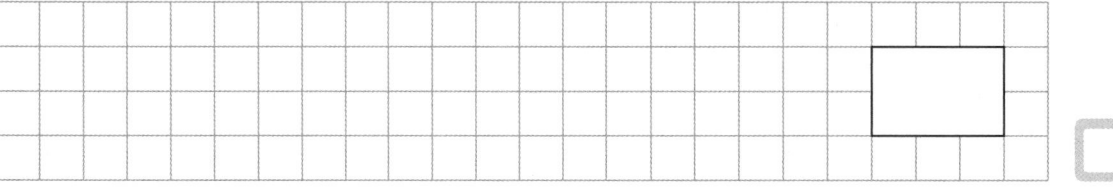

[1]

2 Lauren saved £2.35 of her £5 pocket money. What percentage of her pocket money did she save?

[1]

3 Darius pays £2.16 for a comic that originally cost £2.40. By what percentage has the comic been reduced?

[1]

4 Connor has 80 medals for judo. 35 of his medals are school medals. What percentage of Connor's medals are school medals?

[1]

5 Buggles are $\frac{3}{4}$ the weight of a Tuggle and Tuggles are 40% as heavy as a Ruggle. If a Ruggle weighs 80 kg, how heavy is a Buggle?

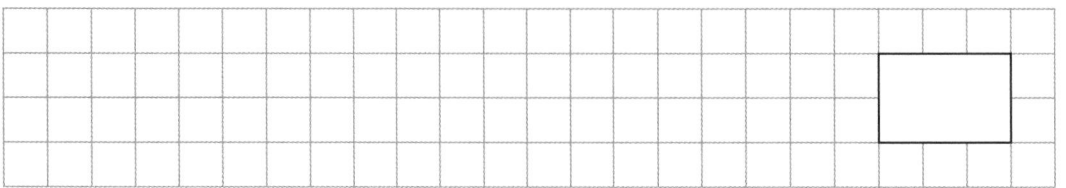

[1]

5

Factors and multiples

> **Helpful Hint** [12]
>
> A **multiple** is a number that another number will divide into exactly.
>
> **Example:** The **multiples** of 3 are 3, 6, 9,... 33, 36, 39...147,150,153... and so on.
>
> A **lowest common multiple (LCM)** is the lowest **multiple** shared by two or more numbers.
>
> **Example:** the LCM of 3 and 4 is 12 because this is the lowest **multiple** shared by both 3 and 4.
>
> **Multiples of 3:** 3, 6, 9, **12**, 15, 18, 21, 24 …
>
> **Multiples of 4:** 4, 8, **12**, 16, 20, 24 …

(A) Write the first five multiples of each number.

1 8 _____ [1]

2 130 _____ [1]

3 24 _____ [1]

4 616 _____ [1]

5 13 _____ [1]

6 214 _____ [1]

(B) Start a list of multiples for each number and then circle the lowest common multiple.

1 4 and 5

   4 _____

   5 _____ [1]

2 4 and 9

   4 _____

   9 _____ [1]

3 2 and 11

   2 _____

   11 _____ [1]

4 5 and 7

   5 _____

   7 _____ [1]

5 6 and 10

   6 _____

   10 _____ [1]

6 6 and 8

   6 _____

   8 _____ [1]

12

💡 Helpful Hint

Factors are numbers that divide exactly into another number.

Example: the factors of 20 are 1, 2, 4, 5, 10, 20.

It helps to list the factors in pairs.

Example: The factors of 144 are 1, 144; 2, 72; 3, 48; 4, 36; 6, 24; 8, 18; 9, 16; 12

The **highest common factor (HCF)** is the largest factor shared by two or more numbers.

Example: the HCF of 10 and 30 is 10 because it is the largest factor shared by 10 and 30.

Factors of 10: 1, 2, 5, **10**

Factors of 30: 1, 2, 3, 5, 6, **10**, 15, 30

Ⓒ List the factors.

1 18 _____ ☐ [1]

2 100 _____ ☐ [1]

3 36 _____ ☐ [1]

4 124 _____ ☐ [1]

5 68 _____ ☐ [1]

6 275 _____ ☐ [1]

Ⓓ List the factors and then circle the highest common factor of these numbers.

1 15 and 50

15 _____

50 _____ ☐ [1]

2 20 and 100

20 _____

100 _____ ☐ [1]

3 75 and 200

75 _____

200 _____ ☐ [1]

4 25 and 70

25 _____

70 _____ ☐ [1]

5 64 and 280

64 _____

280 _____ ☐ [1]

6 250 and 300

250 _____

300 _____ ☐ [1]

☐ 12

Ⓔ Answer these questions as quickly as you can.

1 Write the first 3 multiples of 17 ＿＿＿ ＿＿＿ ＿＿＿ □ [1]

2 Write the first 3 multiples of 27 ＿＿＿ ＿＿＿ ＿＿＿ □ [1]

3 Write the first 3 multiples of 34 ＿＿＿ ＿＿＿ ＿＿＿ □ [1]

4 The factors of 15 are: ＿＿＿ ＿＿＿ ＿＿＿ ＿＿＿ □ [1]

5 The factors of 18 are: ＿＿＿ ＿＿＿ ＿＿＿ ＿＿＿ ＿＿＿ ＿＿＿ □ [1]

6 The factors of 24 are: ＿＿＿ ＿＿＿ ＿＿＿ ＿＿＿ ＿＿＿ ＿＿＿ ＿＿＿ ＿＿＿ □ [1]

7 The HCF of 18 and 20 is ＿＿＿

18 ＿＿＿＿＿＿＿＿＿＿＿＿＿＿＿＿＿＿＿＿＿＿＿＿＿＿＿

20 ＿＿＿＿＿＿＿＿＿＿＿＿＿＿＿＿＿＿＿＿＿＿＿＿＿＿＿ □ [1]

8 The HCF of 40 and 80 is ＿＿＿

40 ＿＿＿＿＿＿＿＿＿＿＿＿＿＿＿＿＿＿＿＿＿＿＿＿＿＿＿

80 ＿＿＿＿＿＿＿＿＿＿＿＿＿＿＿＿＿＿＿＿＿＿＿＿＿＿＿ □ [1]

9 The LCM of 3 and 12 is ＿＿＿

3 ＿＿＿＿＿＿＿＿＿＿＿＿＿＿＿＿＿＿＿＿＿＿＿＿＿＿＿

12 ＿＿＿＿＿＿＿＿＿＿＿＿＿＿＿＿＿＿＿＿＿＿＿＿＿＿＿ □ [1]

10 The LCM of 7 and 9 is ＿＿＿

7 ＿＿＿＿＿＿＿＿＿＿＿＿＿＿＿＿＿＿＿＿＿＿＿＿＿＿＿

9 ＿＿＿＿＿＿＿＿＿＿＿＿＿＿＿＿＿＿＿＿＿＿＿＿＿＿＿ □ [1]

□ / 10

Word problems

(F) Solve the word problems and show your workings out.

1 Mr Cann has a 16 m length of ribbon. He wants to cut it into equal smaller pieces of an exact number of metres. What lengths can he choose to cut the ribbon? _____

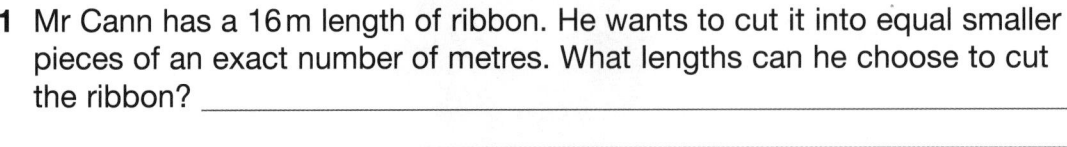

[4]

2 Tom has a bell that rings every 6 seconds and a bulb that lights every 10 seconds. If they ring and light at the same time, how long will it be before they ring and light at the same time again? _____

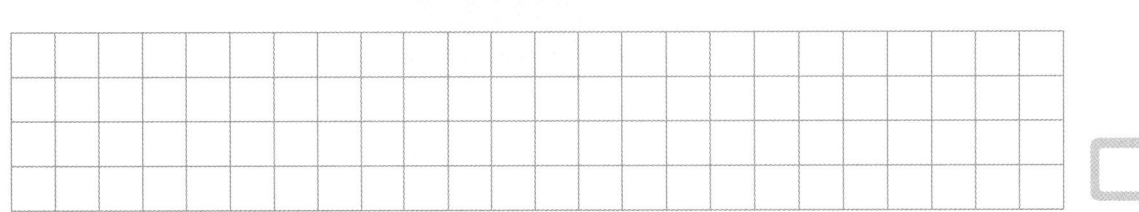

[1]

3 Philip is painting 224 garden gnomes. Every 4th gnome has a blue hat. Every 7th gnome has a fishing rod. Every 8th gnome has a white beard.

  a Which is the first gnome to have a blue hat and a fishing rod? _____

  b Which is the first gnome to have a fishing rod and a white beard? _____

  c Which is the first gnome have a blue hat and a white beard? _____

  d Which is the first gnome to have a blue hat, a fishing rod and

     a white beard? _____

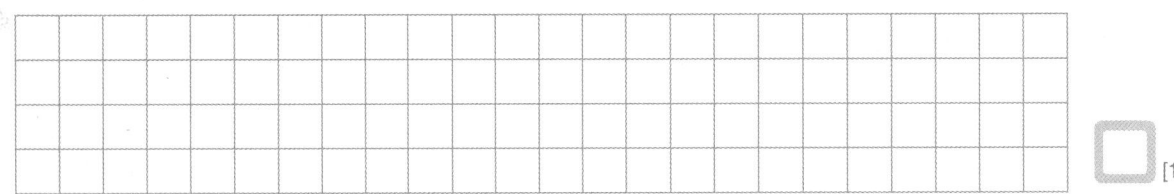

[4]

9

Answers

Unit 1 Number

(A) 1 −80, −30, −10, 90

2 −0.7, −0.2, 0.3, 0.8

3 −4123, −4213, −4231, −4312, −4321

4 −9321, −9312, −931.2, −924.3, −923.4

5 30°C   6 6°C

(B) 1 1081 2 117 3 200 4 24

(C) 1 15 g or 0.015 kg

2 3.3 t

3 1900 g or 1.9 kg

4 3.766 kg

5 452 cl

Word problems

(D) 1 a 112 cakes b £126.26

2 a 19°C   b 9°C

3 a 8 chapters b 4 chapters

Unit 2 Multi-step questions

(A) 1 31   3 20   5 77

2 20   4 8   6 5

(B) 1 58   4 4   7 7.5

2 37   5 34   8 11

3 10   6 3   9 23

(C) 1 67   7 16   13 8

2 137   8 34.75   14 64

3 191   9 16   15 729

4 3077   10 100   16 1000

5 0.9   11 1600

6 2.5   12 2500

(D) 1 £42.67   4 560 g

2 £201.60   5 £17.01
   (20 160 p)

3 46.21 cm

Word problems

(E) 1 a Offer 3 b Offer 2

c £4.12

2 a 14:52   b 37 minutes

3 22 vehicles (7 coaches + 15 minibuses)

Unit 3 Estimating and rounding

(A) 1 700 6<u>0</u>0 500

2 193.472 <u>3</u>00

3 98<u>7</u> 654 321

4 1.234 3<u>2</u>1 47

5 285 9<u>3</u>3 176 000

6 0.283 917 <u>4</u>65

7 9 753 123 49<u>5</u> 000

8 159 357.7<u>4</u>1 268

(B) 1 429 504.312   5 3513.825

2 123.146   6 50 607.082

3 958.585   7 1742.854

4 19 283.747   8 0.123

(C) 1 957 000, 960 000, 1 000 000

2 106 000, 110 000, 100 000

3 859 000, 860 000, 900 000

4 915 285 000, 915 290 000, 915 300 000

5 93 828 000, 93 830 000, 93 800 000

6 8 135 000, 8 130 000, 8 100 000

(D) 1 0.936   0.936 29

2 0.130   0.129 99

3 1.629   1.628 95

4 9.955   9.954 63

5 0.179   0.179 49

6 2.851   2.851 23

(E) 1 600 × 900 = 540 000

2 200 × 34 600 = 6 920 000

(F) 1 14 000 × 81 000 = 1 134 000 000

2 90 000 × 286 000 = 25 740 000 000

(G) 1 2 000 000 × 8 000 000 = 16 000 000

2 9 000 000 × 1 000 000 = 9 000 000

(H) 1 59.9 to 1 d.p. (59.94)

2 17.9 to 1 d.p. (17.86)

Word problems

(I) 1 **a** 20 (allow 19)

    b £20 (allow £19 and £18 if working to 1 decimal place)

2 allow any sensible approximation

    a e.g. $40 \times 80\,mm + 12 \times 100\,mm + 25 \times 200\,mm =$ 9.4 m (actual length 9.11 m to nearest cm)

    b e.g. $15 \times 8\,cm + 20 \times 10\,cm + 10 \times 20\,cm =$ 520 cm (actual length 5.16 m to nearest cm)

3 **a** £9 + £35 + £10 = £54

    b £18 + £28 + £18 = £64

Unit 4 Decimals, fractions and percentages

(A) 1 0.6   60%

2 0.95   95%

3 0.15   15%

4 0.5333   53%

5 0.6153   62%

6 2.75   275%

7 2.8333   283%

8 1.3333   133%

9 8.6666   867%

10 1.7   170%

(B) 1 $\frac{21}{25}$ 3 $2\frac{29}{50}$ 5 $\frac{31}{50}$

2 $\frac{9}{25}$ 4 $\frac{9}{10}$ 6 $4\frac{1}{5}$

(C) 1 £75   3 2460   5 144

2 £129   4 34.3

(D) 1 15%   2 40%   3 35%   4 40%

Word problems

(E) 1 24 stickers   4 43.75%

2 47%   5 24 kg

3 10%

Unit 5 Factors and multiples

(A) 1 8, 16, 24, 32, 40

2 130, 260, 390, 520, 650

3 24, 48, 72, 96, 120

4 616, 1232, 1848, 2464, 3080

5 13, 26, 39, 52, 65

6 214, 428, 642, 856, 1070

(B) 1 4, 8, 12, 16,⑳ 5, 10, 15,⑳

2 4, 8, 12, 16, 20, 24, 28, 32,㊱ 9, 18, 27,㊱

3 2, 4, 6, 8, 10, 12, 14, 16, 18, 20,㉒ 11,㉒

4 5, 10, 15, 20, 25, 30,㉟ 7, 14, 21, 28,㉟

5 6, 12, 18, 24,㉚ 10, 20,㉚

6 6, 12, 18,㉔ 8, 16,㉔

(C) 1 1, 2, 3, 6, 9, 18

2 1, 2, 4, 5, 10, 20, 25, 50, 100

3 1, 2, 3, 4, 6, 9, 12, 18, 36

4 1, 2, 4, 31, 62, 124

5 1, 2, 4, 17, 34, 68

6 1, 5, 11, 25, 55, 275

(D) 1 1, 3,⑤ 15; 1, 2,⑤ 10, 25, 50

2 1, 2, 4, 5, 10,⑳ 1, 2, 4, 5, 10,⑳ 25, 50, 100

3 1, 3, 5, 15,㉕ 75; 1, 2, 4, 5, 8, 10, 20,㉕ 40, 50, 100, 200

4 1,⑤ 25; 1, 2,⑤ 7, 10, 14, 35, 70

5 1, 2, 4,⑧ 16, 32, 64; 1, 2, 4, 5, 7,⑧ 10, 28, 35, 40, 56, 70, 140, 280

6 1, 2, 5, 10, 20,㊿ 125, 250; 1, 2, 3, 4, 5, 6, 10, 12, 15, 20, 25, 30,㊿ 60, 75, 100, 150, 300

(E) 1 17, 34, 51

2 27, 54, 81

3 34, 68, 102

4 1, 3, 5, 15

5 1, 2, 3, 6, 9, 18

6 1, 2, 3, 4, 6, 8, 12, 24

7 2: 1, 2, 3, 6, 9, 18; 1, 2, 4, 5, 10, 20

8 40: 1, 2, 4, 5, 8, 10, 20, 40; 1, 2, 4, 5, 8, 10, 16, 20, 40, 80

9 12: 3, 6, 9, 12,...; 12, 24, ...

10 63: 7, 14, 21, 28, 35, 42, 49, 56, 63, 70,...; 9, 18, 27, 36, 45, 54, 63, 72,...

Word problems

(F) **1** (factors of 16) 1 m, 2 m, 4 m
or 8 m

2 (LCM of 6 and 10)
30 seconds

3 a 28th   **b** 56th

   c 8th   **d** 56th

Unit 6   Mixed operations

(A) **1** $1\frac{1}{2}$ **4** $9\frac{6}{7}$ **7** $3\frac{11}{15}$

2 $14\frac{17}{30}$ **5** $1\frac{9}{10}$

3 $\frac{8}{35}$ **6** $2\frac{19}{28}$

(B) **1** 1324.0   **3** 1660.05

2 £5399.75p   **4** £320.99

(C) **1** $1\frac{19}{30}$ **3** $\frac{4}{7}$ **5** $1\frac{43}{60}$

2 $\frac{1}{40}$ **4** $\frac{33}{40}$

Word problems

(D) **1 a** £128.20   **b** 7 ice creams

2 a £102850   **b** £136893.35

3 a 8 hours   **b** 28 hours

Unit 7   Algebra

(A) **1** ? = 22   **4** $x = 28$

2 ⚽ = 2   **5** $y = 8$

3 ⚽ = 8   **6** $z = 9$

(B) **1** 12 dozen

2 70 ⚑

3 53 ⚽

4 12 gross + 2 score

5 15 dozen + 18 score

6 $52y$

7 $13q$

8 7 thousand

(C) **1** $6n + 7c$ **3** $15m + 22a$

2 $5b + 14d$ **4** $6p + q - 6n$

(D) **1** 3   **2** 6   **3** 5   **4** −3

Word problems

(E) **1** $A = 60$ $B = 20$ $C = 100$

2 ♡ = 10   △ = 15   □ = 20

3 Liam, Connah, Aidan, Niall, Kieran

Unit 8   Sequences

(A) **1** 45, 30 Rule: − 15

2 280, 320 Rule: + 40

3 38, 19 Rule: − 19

4 49, 64 Rule: square numbers

5 36, 45 Rule: +3, +4, +5, etc.

6 192, 384 Rule: doubling

7 108, 165 Rule: + 57

8 8, 4 Rule: halving

(B) **1** −600, −300 Rule: + 50

2 75, 120 Rule: + 15

3 0, −2 Rule: −2

4 −1, 47 Rule: + 12

5 0, −88 Rule: −22

6 169, 225 Rule: squares of odd numbers

7 0.62, 0.38 Rule: −0.06

8 −1, −157 Rule: −52

(C) **1** $2\frac{1}{2}$ Rule: +1

2 1 (allow $\frac{5}{5}$ or equivalent)
Rule: + $\frac{1}{5}$

3 $4\frac{3}{8}$ (allow $\frac{35}{8}$) Rule: + $\frac{7}{8}$

4 $\frac{1}{5}$ Rule: + $\frac{1}{5}$

5 4.75 Rule: − 2.5

6 $\frac{7}{10}$ Rule: − $\frac{1}{10}$

7 0.010101 Rule: ÷ 10

8 − 136.8 Rule: − 37

9 11.4 Rule: + 0.1, + 0.2, + 0.3, etc.

10 54 Rule: 2 independent sequences written alternately 198 −18 each time, 18 +18 each time

11 63 Rule: 2 alternating sequences. (− 28, − 25, − 23, etc.) alternating with (+ 49, + 46, + 43, etc.)

12 64 Rule: 2 independent sequences written alternately, increasing cubic numbers alternating with decreasing square numbers

(D) **1** 12   **5** 96   **9** 5

2 50   **6** 107   **10** 8

3 2   **7** 45

4 8   **8** 7

Word problems

(E) **1** 8

2 **a** 14, 18, 20

    b 116, 332, 764

    c 54, 26, 12

    d 98, 182, 350

Unit 9 Ratio and proportion

(A) **1** 35:44   **7** 3:29:100

2 3:5   **8** $2\frac{1}{8}$

3 1:7   **9** 1:6:42

4 1:5:20   **10** 107:126:149

5 1:17:19   **11** 1:14:42

6 1:12:144   **12** 1:8:24:48

(B) **1** $\frac{2}{7}$ **2** $\frac{13}{27}$ **3** 0.2   **4** 5%

(C) **1** 7, 14, 35   **3** 27, 45, 72

2 36, 48, 60   **4** 72, 84, 96

Word problems

(D) **1** 1:3   **3** 25%

2 $\frac{15}{17}$ **4** 47, 94, 141

Unit 10 Test your skills

(A) **1** −0.8, −0.4 −0.2, 0.6

2 1728   **7** $4\frac{47}{63}$

3 392   **8** $2\frac{1}{8}$

4 99   **9** 1403.1

5 1072   **10** £41.24

6 14.823   **11** 46.187

(B) **1** 0.85   85%

2 36, 72, 108, 144, 180

3 =   **6** 600   **8** 480

4 84   **7** 355   **9** 45

5 0.8

(C) **1** 1, 2, 3, 4, 6, 12

2 2

3 12

4 30   28.5   28.45

5 150   300   45000 (accept any reasonable answer)

6 49.368

7 82<u>5</u>03649

8 ♡ = 18

9 ⚑ = 100

10 63 dozen

11 97 ۞

(D) **1** $10d + 11b$

2 $a = 4, b = 9, c = 6$

3 $a = 8, b = 4, c = 5$

4 $a = 8, b = 7, c = 0$

5 $a = 6, b = 8, c = 6$

6 6

7 74, 69

8 −16, −29 Rule: double the number, add one for the next number, double the number, etc.

9 15   **12** 102

10 31:70   **13** £45.22

11 32, 96, 128

Word problems

(E) **1** £214.60

2 $3\frac{3}{4}$ pizzas

3 984 − 598 ≈ 980 − 600 = 380
309 + 152 ≈ 310 + 150 = 460
23 × 18 ≈ 20 × 20 = 400
339 + 86 ≈ 340 + 90 = 430
460, 430, 400, 380

4 Spencer = 300
Richard = 225
Sam = 150
Lewis = 75

Mixed operations

Helpful Hint

Adding and subtracting fractions

$$\frac{3}{4} + 1\frac{1}{2} - \frac{1}{3} \quad = \quad \frac{9}{12} + \frac{18}{12} - \frac{4}{12} \quad = \quad \frac{23}{12} \quad = \quad 1\frac{11}{12}$$

Step 1 Find the common **denominator**.

Step 2 Change any **mixed numbers** to **improper fractions**.

Step 3 Add and then subtract the **numerators**.

Step 4 Change **improper fractions** back into a mixed number.

Multiplying fractions

$$\frac{3}{4} \times \frac{2}{3} \quad = \quad \frac{3 \times 2}{4 \times 3} \quad = \quad \frac{6}{12} \text{ which in its } \textbf{simplest form} \text{ is } \frac{1}{2}$$

Multiply the **numerators** and multiply the **denominators**. Simplify the fractions where possible.

Dividing fractions

$$\frac{1}{8} \div \frac{1}{3} \quad = \quad \frac{1}{8} \times \frac{3}{1} \quad = \quad \frac{1 \times 3}{8 \times 1} \quad = \quad \frac{3}{8}$$

Remember, to divide a fraction by another fraction, turn the fraction you are dividing by upside down and multiply. So dividing by $\frac{1}{3}$ is the same as multiplying by $\frac{3}{1}$.

To multiply and divide **mixed numbers**, convert the **mixed numbers** to **improper fractions** first.

(A) Calculate and write the answer in its simplest form.

1 $1\frac{2}{7} + \frac{3}{14} =$ _____ [1]

2 $8\frac{1}{2} + 4\frac{2}{3} + 1\frac{2}{5} =$ _____ [1]

3 $\frac{2}{7} \times \frac{4}{5} =$ _____ [1]

4 $3\frac{5}{6} \times 2\frac{4}{7} =$ _____ [1]

5 $2\frac{2}{5} - \frac{1}{2} =$ _____ [1]

6 $5\frac{3}{7} - 2\frac{3}{4} =$ _____ [1]

7 $1\frac{2}{5} \div \frac{3}{8} =$ _____ [1]

7

Helpful Hint

When adding or subtracting, work in columns, keeping the decimal points lined up.

When multiplying decimals, work without the decimal point, then in your answer make sure there are the same number of digits after the decimal point as there are in the question.

When dividing decimals, multiply both the number and **divisor** by 10, 100 or 1000 in order to make the **divisor** a whole number.

```
  237.29
+ 145.10
  382.39
```

```
  64.28
- 21.15
  43.13
```

```
3.04 × 1.2      304
              ×  12
                608
               3040
               3648
              3.648
```

3 digits = 3 **decimal places**

```
52.116 ÷ 0.03  ( × 100)
5211.6 ÷ 3
              1737.2
           3)5211.6

52.116 ÷ 0.03 = 1737.2
```

Remember to estimate the answer before you begin so you can judge whether your answer is reasonable.

(B) Answer these decimal questions and show your workings out.

1
```
  892.4
+ 431.6
```

[1]

2 £215.99 ÷ 0.04

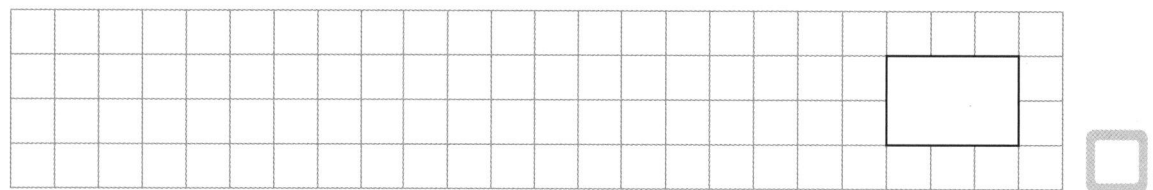

[1]

3
```
  790.5
×   2.1
```

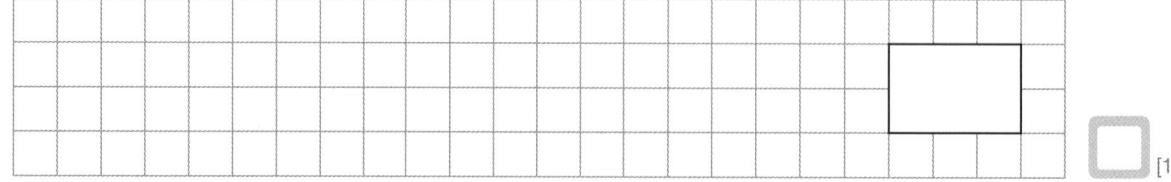

[1]

4
```
  £975.31
- £654.32
```

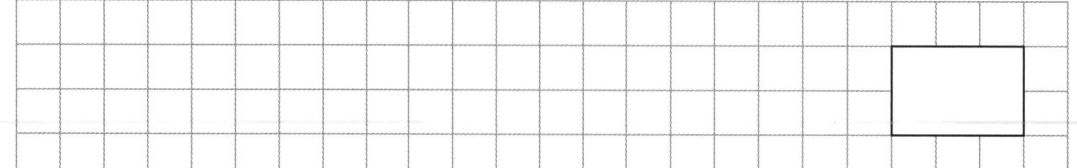

[1]

4

Helpful Hint

Remember to think about what you need to do to the numbers and in which order you need to do the steps. Highlighting numbers and signs can be helpful.

ⓒ Answer these questions as quickly as you can.

1 $\frac{4}{5} + \frac{5}{6} =$ _____

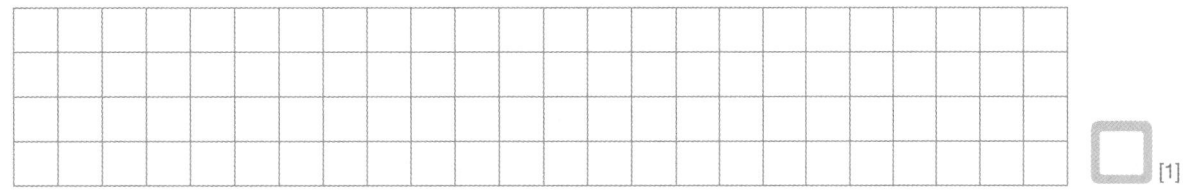

[1]

2 $\frac{9}{10} - \frac{7}{8} =$ _____

[1]

3 $\frac{6}{7} \times \frac{2}{2} =$ _____

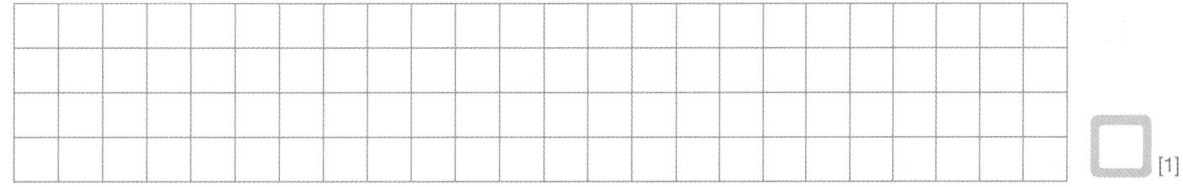

[1]

4 $\frac{3}{4} \div \frac{10}{11} =$ _____

[1]

5 $\frac{11}{12} + \frac{4}{5} =$ _____

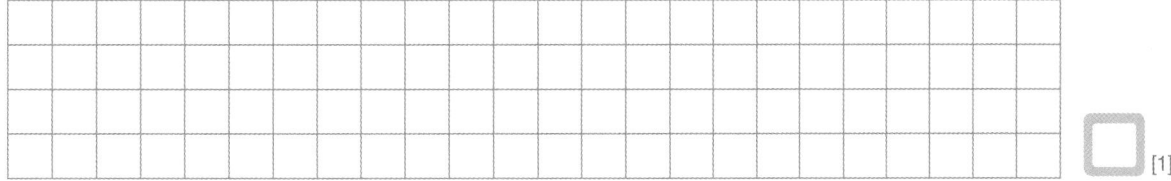

[1]

5

Word problems

(D) Solve the word problems and show your workings out.

1 Dorota is visiting the zoo for her birthday. Her three sisters, brother, mum, dad, grandma and grandpa are all going too.

Zoo Ticket Office			
Tickets		**Refreshments**	
Adults:	£16.30	Ice cream:	£1.25
Children:	£12.60		

a How much does it cost for them all to visit the zoo? _____

The family spend £8.75 on ice cream.

b How many ice creams did they have? _____

[2]

2 In 2011 the value of a flat was £85 000. Each year its value has increased by $\frac{1}{10}$ of the previous year's value.

a What was its value in 2013? _____

b What will its value be in 2016? _____

[2]

3 Phoebe wants to download $2\frac{1}{2}$ hours of music. She can fit 40 hours of tracks on her phone.

a Phoebe has used $\frac{4}{5}$ of the 40 hours. How many hours does she have left? _____

b If she deletes $\frac{5}{8}$ of the music on her phone, how many hours will she then have remaining? _____

[2]

6

Algebra

 Helpful Hint

You have been using **algebra** for a long time – you just haven't called it that before.

Examples: $5 + 7 = ?$ $? = 12$ $10 - \square = 3$ $\square = 7$

             $\circledast - 3 = 15$ $\circledast = 18$ $6 + x = 14$ $x = 8$

Letters are used to stand for unknown numbers because they are a set of symbols that are easy to write.

(A) Find the value of each of these symbols.

1 $29 - 7 = ?$ $? = $ _____ [1]

2 $\circledast + 4 = 6$ $\circledast = $ _____ [1]

3 $7 + \text{⚑} = 15$ $\text{⚑} = $ _____ [1]

4 $23 + 5 = x$ $x = $ _____ [1]

5 $4 + y = 12$ $y = $ _____ [1]

6 $z - 8 = 1$ $z = $ _____ [1]

 Helpful Hint

In **algebra** questions, the answers are not always numerical.

Example: 3 fives + 4 fives = 7 fives

           8 tens – 2 tens = 6 tens

           $7x + 9x = 16x$

           3 gross + 17 dozen + 2 gross – 9 dozen = 5 gross + 8 dozen

(B) Find the value of each of these. Leave your answers in words or symbols.

1 52 dozen – 40 dozen = _____ dozen [1]

2 $17\text{⚑} + 53\text{⚑} = $ _____ ⚑ [1]

3 $74\circledast - 21\circledast = $ _____ $\circledast$ [1]

4 3 gross + 10 score + 9 gross – 8 score = _____ gross + _____ score [2]

5 28 dozen + 3 score – 13 dozen + 15 score = _____ [1]

6 $95y - 43y = $ _____ [1]

7 $7q + 9q - 3q = $ _____ [1]

8 9 thousand + 3 hundred – 2 thousand – 3 hundred = _____ [1]

15

> **Helpful Hint**
>
> In **algebra**, always add or subtract all of the terms that have the same **variables**. (A **variable** is an algebraic symbol, such as a, b, x, y.) We never add or subtract terms that have different **variables**. This is called '**collecting like terms**'.
>
> **Example:** $3a + 2b - a + 3b = 2a + 5b$
>
> There are also negative answers in **algebra**.
>
> **Example:** $2a + 3b - 3a - 5b = -a - 2b$
>
> There may be brackets in an **expression**. To remove the brackets multiply the number outside the brackets with everything inside the brackets.
>
> **Example:** $3(a + 2b)$ means $3 \times a + 3 \times 2b$. This gives us $3a + 6b$.
>
> Remember, $3x$ means 3 lots of x or $3 \times x$. If there is only one of the unknown number use y, not $1y$.

(c) Collect like terms.

1 $5n + 3c + 2n + 4c - n =$ _____

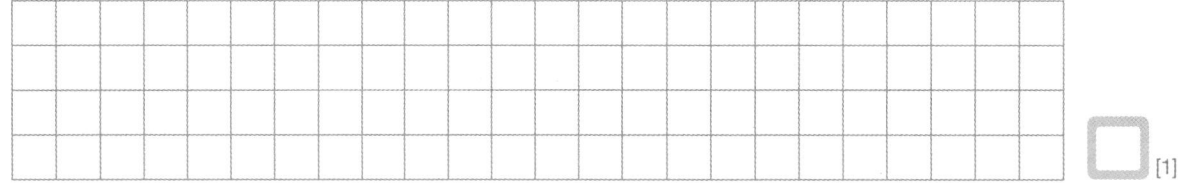

[1]

2 $4b + 2d + 2(b + 6d) - b =$ _____

[1]

3 $3(m + 10a) + 4(3m - 2a) =$ _____

[1]

4 $2p + 3q - 5n + 4p - n - 2q =$ _____

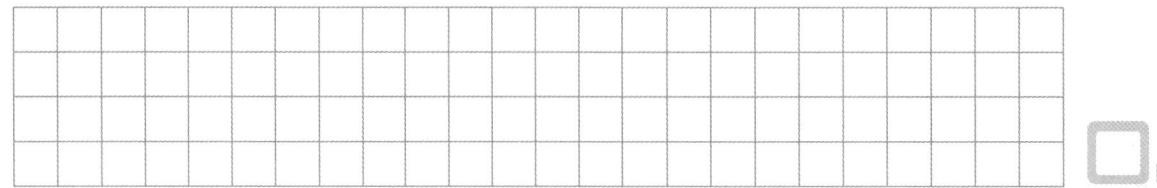

[1]

4

Helpful Hint

Solve an **equation** to find out what value a letter represents. Ensure the terms with letters are on one side and the numbers are on the other side. To do this, rearrange the terms like this:

$3a + 2 = a + 8$ Subtract '2' from both sides to leave '$3a$' on the left side.

$3a + 2 - 2 = a + 8 - 2$ **Collect like terms.**

$3a = a + 6$ Subtract 'a' from both sides to leave '6' on the right side.

$3a - a = a + 6 - a$ **Collect like terms.**

$2a = 6$ Now divide both sides by 2 to simplify the numbers.

$a = 3$

Remember that every action must be done on both sides of the equals sign.

Ⓓ Find the number that y represents.

1 $6y + 3 = 2y + 15$

[1]

2 $4y + 1 = y + 19$

[1]

3 $5y - 12 = y + 8$

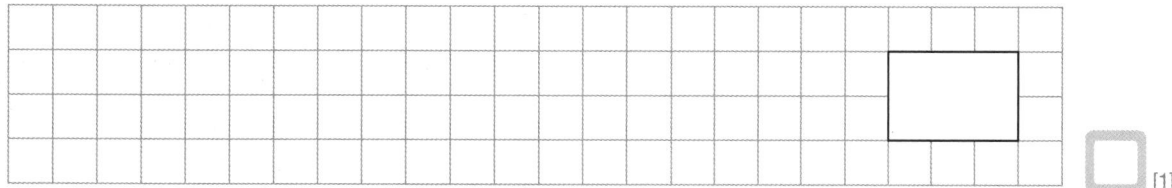

[1]

4 $2(3y + 6) = 3(y + 1)$

[1]

4

Word problems

 Solve the word problems and show your workings out.

1 Sophia has a huge box full of fruit. She has apples (A), bananas (B) and cherries (C). Sophia says: "Can you find how many apples, bananas and cherries I have?" A, B and C each stand for a whole number. $A + B + C = 180$ pieces of fruit. A is three times bigger than B. C is five times bigger than B.

$A =$ _____ $B =$ _____ $C =$ _____

[3]

2 Find the value of each shape.

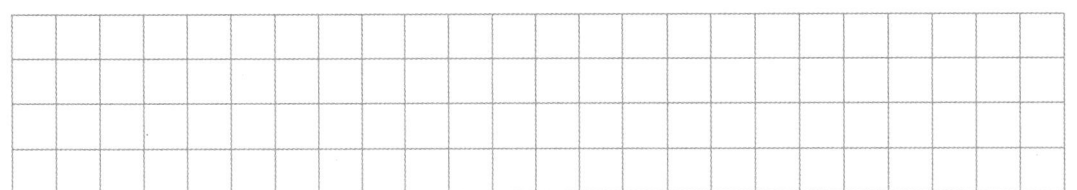

$\heartsuit + \heartsuit + \triangle = 35$ $\triangle + \square + \heartsuit = 45$

$\square + \square + \square = 60$ $\heartsuit + \heartsuit + \square = 40$

$\heartsuit =$ _____ $\triangle =$ _____ $\square =$ _____

[3]

3 The expressions below show the ages of five brothers where $x = 5$ and $y = 2$. For example, Connah's age is shown by the expression $5x - 10y$. The youngest brother has the smallest result. The eldest brother has the largest result. Place the brothers in order of age from youngest to eldest.

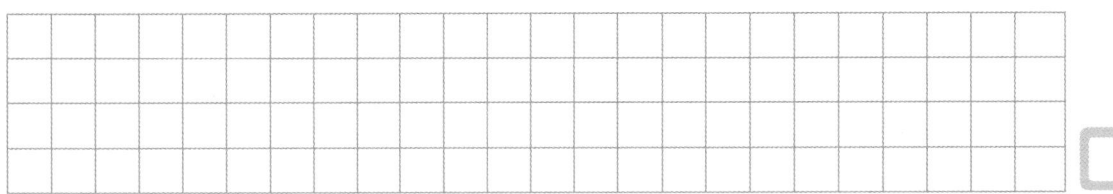

Connah: $5x - 10y$ Aidan: $2x + 3y$ Liam: $x - y$

Niall: $3x + y$ Kieran: $x + 8y$

_____   _____   _____   _____   _____

[5]

11

Sequences

  Helpful Hint

Number sequences reminders

A square number is the product of a number multiplied by itself.

A cube number is the product of a number multiplied by itself, then by itself again.

A prime number has only two factors, one and itself.

Triangle numbers can be represented by a triangle of dots. Each triangle number adds one extra row to the triangle each time.

Example:

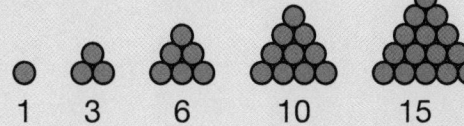

1   3   6   10   15

Ⓐ Write the next two numbers in each sequence.

1 135   120   105   90   75   60   _____   _____   [2]

2 40   80   120   160   200   240   _____   _____   [2]

3 152   133   114   95   76   57   _____   _____   [2]

4 1   4   9   16   25   36   _____   _____   [2]

5 3   6   10   15   21   28   _____   _____   [2]

6 3   6   12   24   48   96   _____   _____   [2]

7 −234   −177   −120   −63   −6   51   _____   _____   [2]

8 512   256   128   64   32   16   _____   _____   [2]

Ⓑ Complete these sequences.

1 _____   −550   −500   −450   −400   −350   _____   −250   [2]

2 15   30   45   60   _____   90   105   _____   [2]

3 10   8   6   4   2   _____   _____   −4   [2]

4 −37   −25   −13   _____   11   23   35   _____   [2]

5 66   44   22   _____   −22   −44   −66   _____   [2]

6 1   9   25   49   81   121   _____   _____   [2]

7 _____   0.56   0.5   0.44   _____   0.32   0.26   0.2   [2]

8 51   _____   −53   −105   _____   −209   −261   −313   [2]

32

Helpful Hint

If there are fractions in a **number sequence**, you may need to use **equivalent fractions** to work out the difference between them.

Sometimes two **sequences** can run alternately. If a **sequence** doesn't seem to make sense, look at every other number and see if this works.

ⓒ Complete these sequences.

1 $-1\frac{1}{2}$ $-\frac{1}{2}$ $\frac{1}{2}$ $1\frac{1}{2}$ _____ $3\frac{1}{2}$ [1]

2 $\frac{1}{5}$ $\frac{4}{10}$ $\frac{9}{15}$ $\frac{4}{5}$ _____ $1\frac{2}{10}$ [1]

3 $1\frac{3}{4}$ $2\frac{5}{8}$ $3\frac{1}{2}$ _____ $5\frac{1}{4}$ $6\frac{1}{8}$ [1]

4 _____ $\frac{4}{10}$ $\frac{3}{5}$ $\frac{8}{10}$ 1 $1\frac{4}{20}$ [1]

5 9.75   7.25   _____   2.25 -0.25 -2.75 [1]

6 1 $\frac{9}{10}$ $\frac{4}{5}$ _____ $\frac{3}{5}$ $\frac{1}{2}$ [1]

7 10 101   1010.1   101.01   10.101   1.0101   0.101 01   _____   [1]

8 85.2   48.2   11.2 -25.8 -62.8 -99.8 _____   [1]

9 11.3   _____   11.6   11.9   12.3   12.8   [1]

10 198   18   180   36   162   _____   144   72   126   90   [1]

11 52   14   _____   28   74   42   85   56   96   70   [1]

12 1   121   8   100   27   81   64   _____   125   49   216   [1]

Helpful Hint

Solve these types of number problems by reversing the operations.

A **number machine** has a series of rules. A number is put in and it follows each rule, and then the output is given like this:

$3 \Rightarrow \boxed{\times 2} \Rightarrow \boxed{+ 5} = 11$

If we don't know the input number is 3 we can simply use the **inverse operations** in reverse order like this:

$11 \Rightarrow \boxed{- 5} \Rightarrow \boxed{\div 2} = 3$

Example:

$n \Rightarrow \boxed{\times 2} \Rightarrow \boxed{+ 7} \boxed{\times 2} = 46$ What is n?

$46 \Rightarrow \boxed{\div 2} \Rightarrow \boxed{- 7} \boxed{\div 2} = n$ $n = 8$

Remember to check your answer by substituting the **value of n** into the problem.

(D) Find the value of n.

1 $n \Rightarrow \boxed{\times 3} \Rightarrow \boxed{+ 6} \Rightarrow \boxed{\div 2} \Rightarrow \boxed{+ 8} = 29$ $n = $ ⬚ ⬚ [1]

2 $n \Rightarrow \boxed{- 10} \Rightarrow \boxed{\div 2} \Rightarrow \boxed{+ 1} \Rightarrow \boxed{\div 3} = 7$ $n = $ ⬚ ⬚ [1]

3 $n \Rightarrow \boxed{\times 2} \Rightarrow \boxed{\times 3} \Rightarrow \boxed{\times 4} \Rightarrow \boxed{+ 2} = 50$ $n = $ ⬚ ⬚ [1]

4 $n \Rightarrow \boxed{+ 10} \Rightarrow \boxed{\div 9} \Rightarrow \boxed{+ 8} \Rightarrow \boxed{- 7} = 3$ $n = $ ⬚ ⬚ [1]

5 $n \Rightarrow \boxed{\div 8} \Rightarrow \boxed{\div 3} \Rightarrow \boxed{\times 2} \Rightarrow \boxed{+ 2} = 10$ $n = $ ⬚ ⬚ [1]

6 $n \Rightarrow \boxed{+ 4} \Rightarrow \boxed{\times 3} \Rightarrow \boxed{+ 4} \Rightarrow \boxed{\times 3} = 1011$ $n = $ ⬚ ⬚ [1]

7 $n \Rightarrow \boxed{- 5} \Rightarrow \boxed{\div 2} \Rightarrow \boxed{- 5} \Rightarrow \boxed{\div 3} = 5$ $n = $ ⬚ ⬚ [1]

8 $n^2 \Rightarrow \boxed{+ 3} \Rightarrow \boxed{\div 2} \Rightarrow \boxed{+ 4} \Rightarrow \boxed{\div 5} = 6$ $n = $ ⬚ ⬚ [1]

9 $n^3 \Rightarrow \boxed{- 50} \Rightarrow \boxed{+ 5} \Rightarrow \boxed{\div 2} = 40$ $n = $ ⬚ ⬚ [1]

10 $n^2 \Rightarrow \boxed{+ 8} \Rightarrow \boxed{\div 2} \Rightarrow \boxed{+ 8} \Rightarrow \boxed{\div 2} = 22$ $n = $ ⬚ ⬚ [1]

⬚ 10

Word problems

(E) Solve the word problems and show your workings out.

1 Amelia starts with her own age, doubles it, adds seven and then doubles it again to find her dad's age. Her dad is 46, so how old is Amelia?

[1]

2 Write the next three numbers for each of these sequences.

a Add 22 to the previous number and then halve it.

6 _____ _____ _____

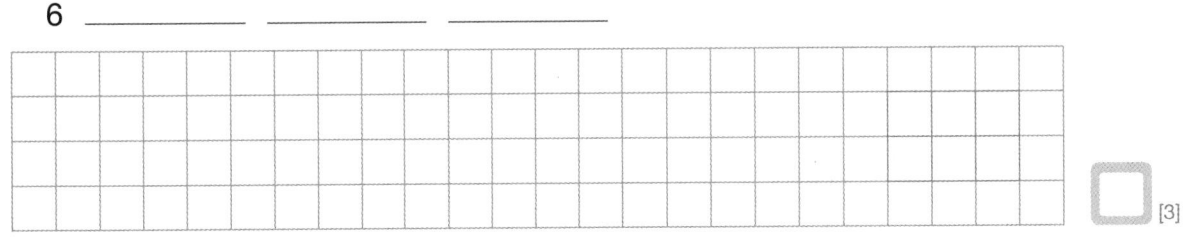

[3]

b Add 50 to the previous number and then double it.

8 _____ _____ _____

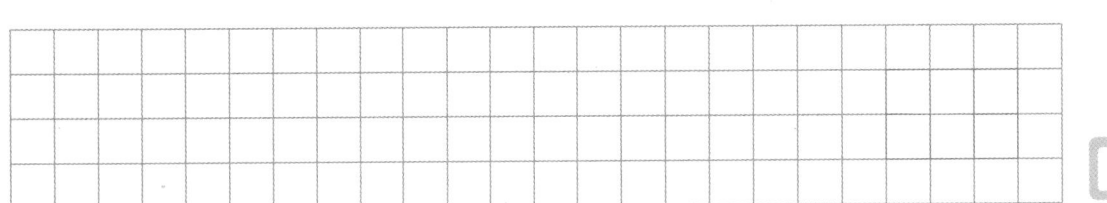

[3]

c Subtract 2 from the previous number and then halve it.

110 _____ _____ _____

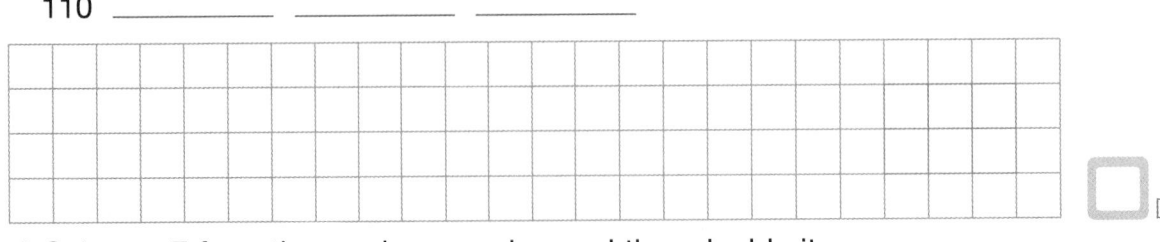

[3]

d Subtract 7 from the previous number and then double it.

56 _____ _____ _____

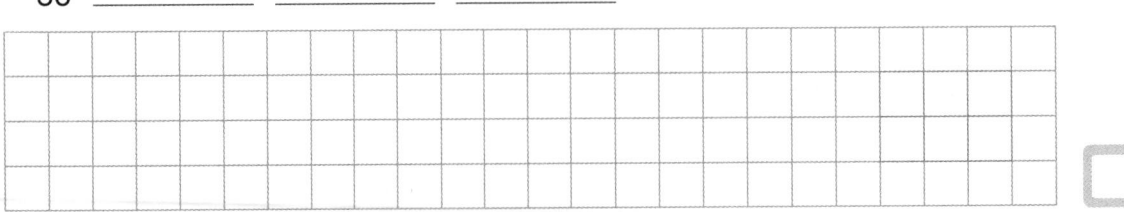

[3]

13

Ratio and proportion

Helpful Hint

A **ratio** is a way of comparing one quantity to another. It is best to show a **ratio** in its **simplest form** (as with fractions). You can simplify the amounts by dividing by common **factors**.

Example: Polly's Pet Shop has 7 black puppies and 14 brown puppies. The **ratio** of black puppies to brown puppies is 7:14, which when simplified (by dividing by 7) is 1:2.

This means that for every 1 black puppy in the pet shop, there will be 2 brown puppies.

There can be more than 2 amounts in a **ratio**. For example, Lukas, Aleksander and Nikolas have 60 books in the **ratio** 10:20:30.

Then put the **ratios** in their **simplest form**, 1:2:3.

Ⓐ Write these ratios in their simplest forms.

1 175:220 = _____ ☐ [1]

2 48:80 = _____ ☐ [1]

3 23:161 = _____ ☐ [1]

4 13:65:260 = _____ ☐ [1]

5 7:119:833 = _____ ☐ [1]

6 12:144:1728 = _____ ☐ [1]

7 15:145:500 = _____ ☐ [1]

8 6:15:33 = _____ ☐ [1]

9 120:720:5040 = _____ ☐ [1]

10 214:252:298 = _____ ☐ [1]

11 49:686:2058 = _____ ☐ [1]

12 19:152:456:912 = _____ ☐ [1]

☐ 12

Unit 9

> **Helpful Hint**
>
> **Proportion** tells us what fraction of a whole thing needs to be considered.
>
> **Example:** Polly's Pet Shop has 56 animals and 21 of them are dogs. What **proportion** of the animals are dogs?
>
> The **proportion** is 21 out of 56 or $\frac{21}{56}$. To simplify, simply divide both the **numerator** and the **denominator** by 7 to give the fraction $\frac{3}{8}$.
>
> $\frac{3}{8}$ of the animals are dogs.
>
> We can also write this as a decimal (0.375) or a **percentage** (37.5%).
>
> 37.5% of the animals are dogs.

(B) Write these proportions in their simplest forms.

1 12 out of 42 as a fraction.

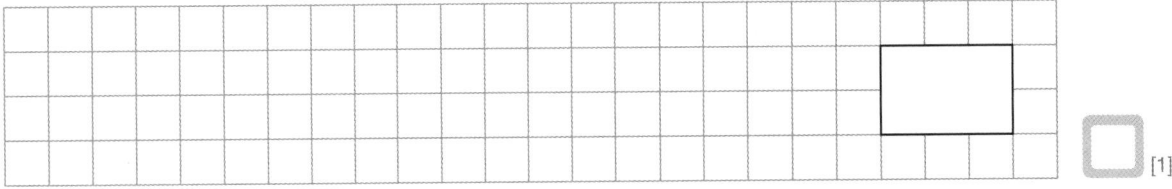

[1]

2 52 out of 108 as a fraction.

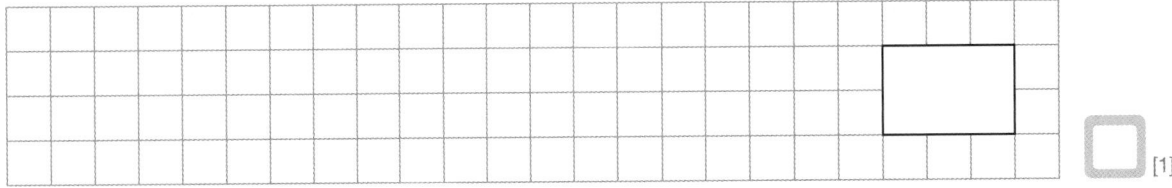

[1]

3 65 out of 325 as a decimal.

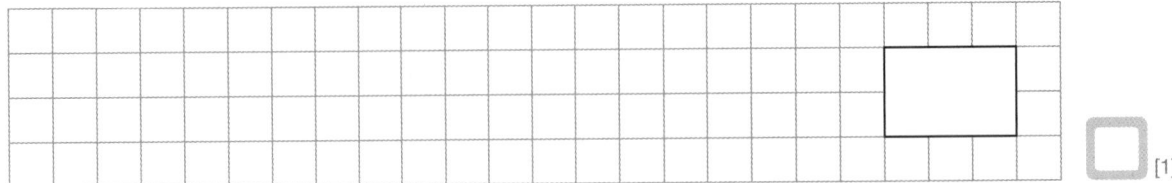

[1]

4 120 out of 2400 as a percentage.

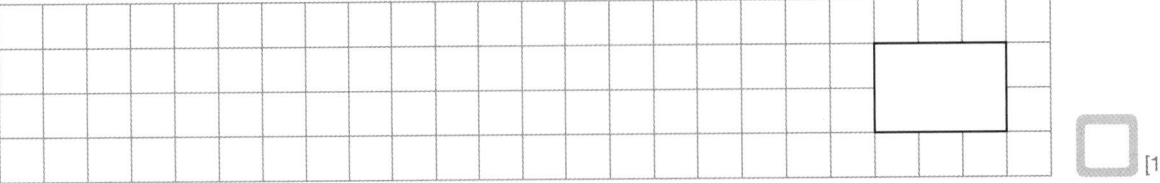

[1]

4

Helpful Hint

Use the following technique to solve a **ratio** problem.

Example: I have 50 red, blue and green counters in the **ratio** 2:3:5. How many of each colour do I have?

Step 1 Add the number of parts together. (2 + 3 + 5 = 10)

Step 2 Divide this number into the total number of counters to find how many counters there are in **one** part. (50 ÷ 10 = 5)

Step 3 Multiply this by each number in the **ratio**. (5 × 2 = 10 red counters) (5 × 3 = 15 blue counters) (5 × 5 = 25 green counters).

We can check our results by adding together 10 + 15 + 25 = 50 counters.

ⓒ Write how many red, blue and green counters there are in these ratios.

1 56 in the ratio 1:2:5.   Red = _____ Blue = _____ Green = _____

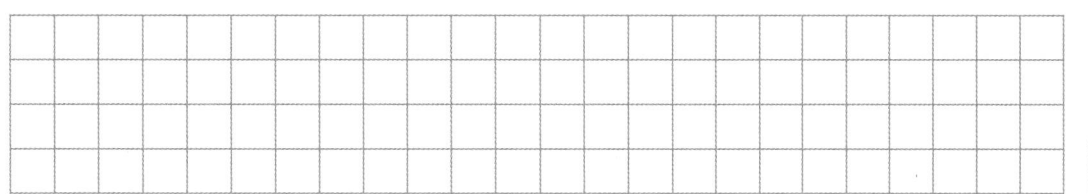

[3]

2 144 in the ratio 6:8:10.   Red = _____ Blue = _____ Green = _____

[3]

3 144 in the ratio 3:5:8.   Red = _____ Blue = _____ Green = _____

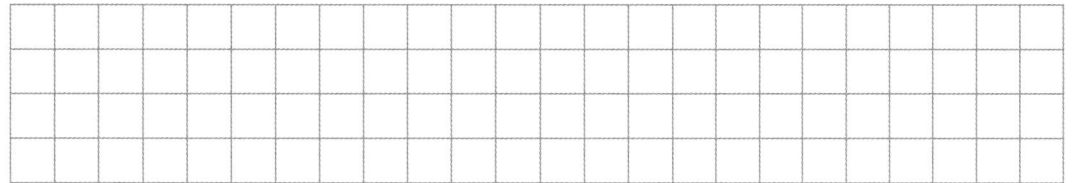

[3]

4 252 in the ratio 12:14:16.   Red = _____ Blue = _____ Green = _____

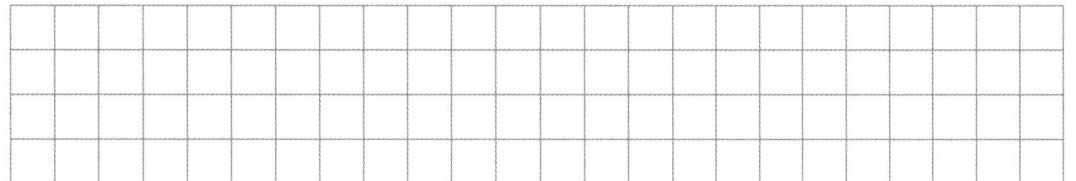

[3]

12

Unit 9

Word problems

Ⓓ Solve the word problems and show your workings out.

1 A builder uses 12 buckets of cement to 36 buckets of sand.

What is the ratio of cement to sand in its simplest form?

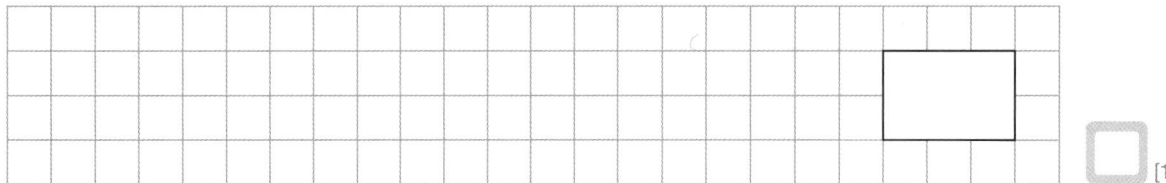

[1]

2 Seisdon High school has 450 boys and 510 girls.

What is the proportion of boys to girls as a fraction in its simplest form?

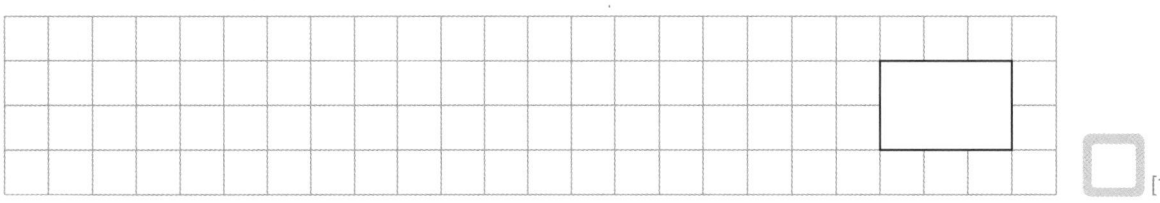

[1]

3 Jonah has 128 Franmere stickers in his collection of 512 football stickers.

What proportion of Jonah's collection are Franmere stickers? Write your answer as a percentage to the nearest whole number.

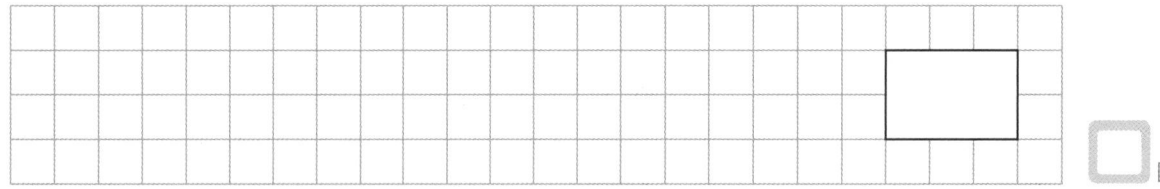

[1]

4 Rushda made 282 snacks. She made bowls of kichuri, samosas and pakoras in the ratio 1:2:3. How many of each does Rushda make?

kichuri = _____   samosas = _____   pakoras = _____

[3]

6

Test your skills

(A) Answer these questions. You do not need to show your workings out.

1 Look at the arrows and fill in the numbers on number line.

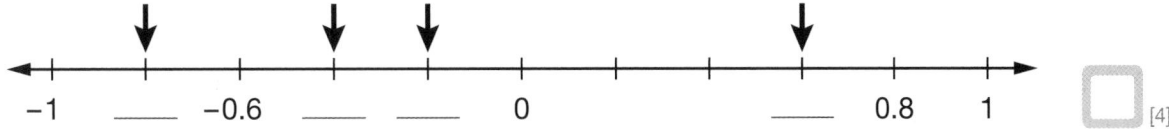

$\quad$ [4]

2 $8^2 \times 3^3 =$ _____ [1]

3 $2^3 \times 7^2 =$ _____ [1]

4 $\sqrt{81} \times \sqrt{121} =$ _____ [1]

5 $10.72\,\text{l} =$ _____ cl [1]

6 $14\,823\,000\,\text{mm} =$ _____ km [1]

7 $2\frac{6}{7} + 1\frac{8}{9} =$ _____ [1]

8 $3\frac{7}{8} - 1\frac{3}{4} =$ _____ [1]

9
```
   876.3
 +526.8
 _____
```
[1]

10
```
  £215.99
 -£174.75
 _____
```
[1]

11
```
   61.043
 -14.856
 _____
```
[1]

(B) Answer these questions.

1 Write $\frac{17}{20}$ as a decimal then as a percentage. _____ _____ [2]

2 Write the first 5 multiples of 36 = _____ _____ _____ _____ _____ [5]

3 Use <, = or > in the correct place. $\frac{11}{12}$ of 276 _____ 46% of 550 [1]

4 $\frac{7}{10}$ of 120 _____ [1]

5 0.02 of 40 _____ [1]

6 $\frac{3}{4}$ of 800 _____ [1]

7 0.25 of 1420 _____ [1]

8 80% of 600 _____ [1]

9 30% of 150 _____ [1]

ⓒ Answer these questions. You do not need to show your workings out.

1 The factors of 12 are: _____ ☐ [1]

2 The HCF of 30 and 46 is _____

30 _____

46 _____ ☐ [1]

3 The LCM of 4 and 6 is _____

4 _____

6 _____ ☐ [1]

4 Round 28.452 to the nearest ten _____, tenth _____, hundredth _____ ☐ [3]

5 Estimate the answer: $149 \times 314 \approx$ _____ × _____ = _____ ☐ [1]

6 Write the number 49.367 81 to 3 decimal places _____ ☐ [1]

7 Underline the digit that is in the 100 000 place in the number 82 503 649. ☐ [1]

8 Find the number that ♡ represents. $32 + ♡ - 14 = 36$ ♡ = _____ ☐ [1]

9 Find the number that ℗ represents. $℗ - 93 + 21 = 28$ ℗ = _____ ☐ [1]

Find the value of each of these. Leave your answers in words or symbols.

10 76 dozen – 13 dozen = _____ dozen ☐ [1]

11 $36✿ + 61✿ =$ _____ ✿ ☐ [1]

13

⒟ Answer these questions.

1 Collect like terms. $6d + 3b + 4(d + 2b) =$ _____ [1]

Work out these column addition and subtraction questions to find out the value of the letters.

2
```
  4 a 1
+ 1 6 b
─────────
  c₁1₁0
```

3
```
  a 9 6
− 5 b 1
─────────
  3 5 c
```

4
```
  2 a 1
+ b 8 9
─────────
 10₁7₁c
```

5
```
  9 2 0
− 2 b a
─────────
  c 3 4
```

$a =$ _____ $a =$ _____ $a =$ _____ $a =$ _____

$b =$ _____ $b =$ _____ $b =$ _____ $b =$ _____

$c =$ _____ $c =$ _____ $c =$ _____ $c =$ _____ [12]

6 Find the number that y represents. $7y + 12 = 3y + 36$ $y =$ _____ [1]

7 Write the next two numbers in this sequence.

146   129   114   101   90   81   _____   _____ [2]

8 Complete this sequence.

$-4\frac{1}{2}$ −9   −8   _____   −15   −30   _____   −58   −57 [2]

9 Find the value of n.

n $\boxed{\times 4}$  $\boxed{-6}$ $\boxed{\div 2}$ 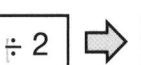 $\boxed{+14}$ $= 41$ $n =$ $\boxed{}$ [1]

10 Write this ratio in its simplest form. $186 : 420 =$ _____ : _____ [1]

11 I have some red, blue and green counters in the ratio 1:3:4. If there are 256 counters altogether, find how many red, blue and green counters there are.

Red = _____ Blue = _____ Green = _____ [3]

12 $6(8 − 3) + 4(9 \times 2) =$ _____ [1]

13 £60 − (14p + 86p + £13.78) = _____ [1]

25

Unit 10

Word problems

(E) Solve the word problems and show your workings out.

1 Ticket prices for the pantomime are £18.50 for an adult and £12.95 for a child. If 6 adults and 8 children visit the pantomime, how much would the total ticket price cost? _____

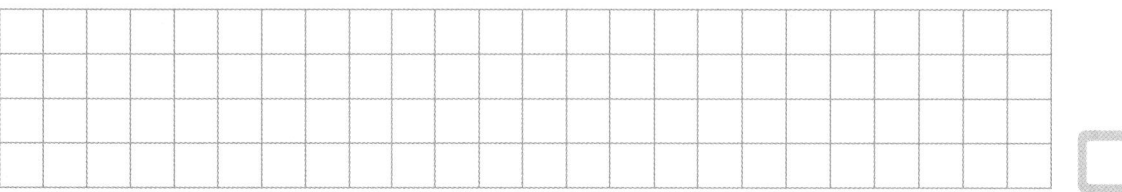

[1]

2 Andrew and John bake 12 mini pizzas and eat $\frac{1}{3}$ of them. Steven eats $1\frac{1}{4}$ pizzas. Nick eats $2\frac{1}{3}$ pizzas and Chris eats $\frac{2}{3}$ of a pizza.

How much pizza is left? _____

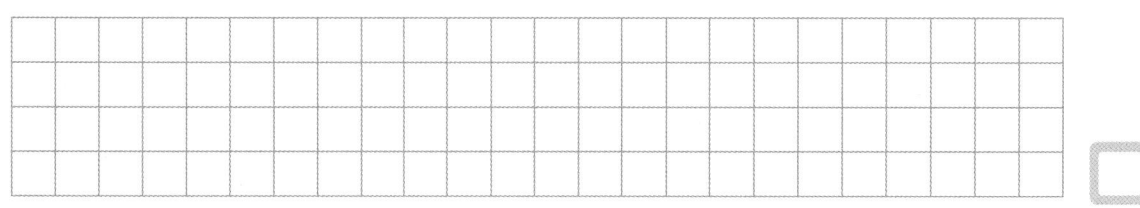

[1]

3 Round the numbers up or down to the nearest ten. Answer the problems and place the answers in order, largest estimated answer first.

$984 - 598 \approx$ ☐ $\quad 309 + 152 \approx$ ☐ $\quad 23 \times 18 \approx$ ☐ $\quad 339 + 86 \approx$ ☐

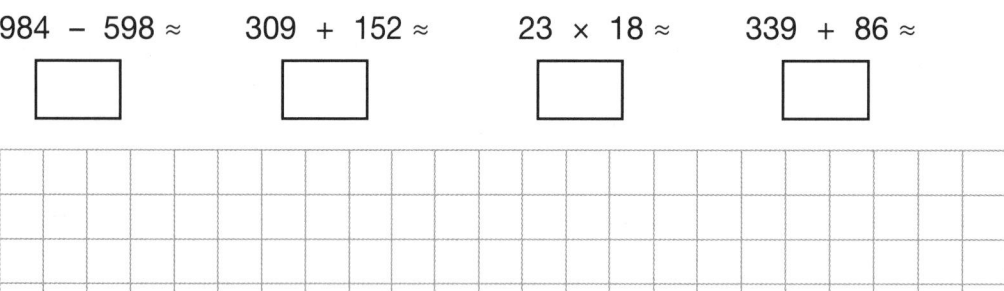

[4]

4 Spencer, Richard, Sam and Lewis have 750 books in the ratio 4:3:2:1. How many books do they have each?

Spencer = _____ Richard = _____

Sam = _____ Lewis = _____

[4]

10

Key words

Algebra using letters to represent numbers

BODMAS an order of working out which parts of a calculation we do first. BODMAS stands for Brackets, Order, Division, Multiplication, Addition, Subtraction

Collecting like terms adding or subtracting terms with the same letters in algebra, for example $6a - a = 5a$

Decimal fraction a fraction with a multiple of 10 as the denominator that can then be placed on the decimal system using a decimal point, for example $\frac{40}{100} = 0.40$

Decimal place the place value of a digit after the decimal point, for example 2 decimal places means the number is written with digits to the hundredths place

Denominator the bottom number of a fraction is called the denominator. It tells you how many equal parts the shape or amount has been divided into

Divisor a number that is divided into another, for example $28 \div \mathbf{7}$

Equation a statement with an equals sign (=) showing that two expressions have the same value, for example $3 + 2 = 5$ or $x - 9 = 5 + 6$

Equivalent fractions fractions with the same value that use different numbers, for example $\frac{2}{3}$ and $\frac{4}{6}$

Estimation (≈) an approximate amount calculated by rounding up or down, for example $98 \times 41 \approx 4000$ (100×40)

Expression a group of numbers or algebraic letters, linked by signs for operations (such as +, ÷) and usually not including an equals sign, for example $x + y$ or $4(x - y)$

Factor a number that divides exactly into a larger number

Greater than (>) how we show that one number is larger than another, for example $47 > 35$

Highest common factor (HCF) the highest number that will divide exactly into two or more numbers, for example the HCF of 8 and 12 is 4 *(factors of 8 = 1, 2, **4**, 8) (factors of 12 = 1, 2, 3, **4**, 6, 12)*

Improper fraction a fraction where the numerator is bigger than the denominator

Inverse operation (or reverse operation) the inverse of addition is subtraction. The inverse of multiplication is division. We can use inverse operations to check an answer

Less than (<) how we show that one number is smaller than another, for example $35 < 47$

Lowest common multiple (LCM) the lowest multiple of two or more numbers, for example the LCM of 2 and 3 is 6 *(multiples of 2 = 2, 4, **6**, 8, ...) (multiples of 3 = 3, **6**, 9, ...)*

Mixed number a number with both a whole number and a fraction, for example $2\frac{1}{2}$

Multiple the result of one number multiplied by another, for example $1 \times 3 = \mathbf{3}, 2 \times 3 = \mathbf{6}$

Multi-step problem solving a problem that needs more than one step to solve it. You often need to use more than one operation

Number sequence a series of numbers that form a pattern, for example *10, 11, 12, 13* or *10, 20, 30, 40*

Numerator the top number of a fraction that tells you how many parts of the total amount have been taken, for example $\frac{2}{3}$ *(**two** thirds)* or $\frac{1}{2}$ *(**one** half)*

Output (or number) machine a fixed operation, or operations, that a number must follow in order to find an end result, for example if the input is '2' and the output machine is $\boxed{+5}$, the output is '7'

Percentages (%) percentages tell you a number out of 100, for example *52% = 52 out of 100*

Place value the place value tells you the value of a digit in a number, for example units or tens

Proportion proportion tells us how much out of the whole amount we have, for example if a school of 78 pupils has 38 boys, the proportion of boys is $\frac{38}{70}$

Ratio ratio tells us how much we have in relation to another amount, for example if a school of 78 pupils has 38 boys, the ratio of boys to girls is 38:40 or 19:20

Rounding to approximate a number to a specified level of accuracy, for example to the nearest 100 or to 3 decimal places

Simplest form reducing the numbers in a fraction or ratio until they can no longer be divided by a common factor, for example 15:10 in its simplest form is 3:2 (the common factor is 5)

Value of n n represents a number, for example if $n = 3$, then $4n = 12$ but if $n = 10$ then $4n = 40$

Variable an algebraic symbol or letter such as a or x that represents an unknown number

Progress chart

How did you do? Fill in your score. Shade the matching boxes so that
you can see how well you are doing in the different units.

50%   100%   50%   100%

Unit 1, p3 Score: __ / 12   **Unit 6**, p27 Score: __ / 7

Unit 1, p4 Score: __ / 4   **Unit 6**, p28 Score: __ / 4

Unit 1, p5 Score: __ / 5   **Unit 6**, p29 Score: __ / 5

Unit 1, p6 Score: __ / 6   **Unit 6**, p30 Score: __ / 6

Unit 2, p7 Score: __ / 15   **Unit 7**, p31 Score: __ / 15

Unit 2, p8 Score: __ / 16   **Unit 7**, p32 Score: __ / 4

Unit 2, p9 Score: __ / 10   **Unit 7**, p33 Score: __ / 4

Unit 2, p10 Score: __ / 6   **Unit 7**, p34 Score: __ / 11

Unit 3, p11 Score: __ / 16   **Unit 8**, p35 Score: __ / 32

Unit 3, p12 Score: __ / 30   **Unit 8**, p36 Score: __ / 12

Unit 3, p13 Score: __ / 8   **Unit 8**, p37 Score: __ / 10

Unit 3, p14 Score: __ / 6   **Unit 8**, p38 Score: __ / 13

Unit 4, p15 Score: __ / 26   **Unit 9**, p39 Score: __ / 12

Unit 4, p16 Score: __ / 5   **Unit 9**, p40 Score: __ / 4

Unit 4, p17 Score: __ / 4   **Unit 9**, p41 Score: __ / 12

Unit 4, p18 Score: __ / 5   **Unit 9**, p42 Score: __ / 6

Unit 5, p19 Score: __ / 12   **Unit 10**, p43 Score: __ / 28

Unit 5, p20 Score: __ / 12   **Unit 10**, p44 Score: __ / 11

Unit 5, p21 Score: __ / 10   **Unit 10**, p45 Score: __ / 25

Unit 5, p22 Score: __ / 9   **Unit 10**, p46 Score: __ / 10